Artificial Intelligence Methods Applied to Urban Remote Sensing and GIS

Artificial Intelligence Methods Applied to Urban Remote Sensing and GIS

Editors

Chang-Wook Lee
Hyangsun Han
Hoonyol Lee
Yu-Chul Park

MDPI • Basel • Beijing • Wuhan • Barcelona • Belgrade • Manchester • Tokyo • Cluj • Tianjin

Editors
Chang-Wook Lee
Science edcation
Kangwon National University
Chuncheon
Korea, South

Hyangsun Han
Geophysics
Kangwon National University
Chuncheon
Korea, South

Hoonyol Lee
Geophysics
Kangwon National University
Chuncheon
Korea, South

Yu-Chul Park
Geophysics
Kangwon National University
Chuncheon
Korea, South

Editorial Office
MDPI
St. Alban-Anlage 66
4052 Basel, Switzerland

This is a reprint of articles from the Special Issue published online in the open access journal *Remote Sensing* (ISSN 2072-4292) (available at: www.mdpi.com/journal/remotesensing/special_issues/AI_urban_remote_sensing).

For citation purposes, cite each article independently as indicated on the article page online and as indicated below:

LastName, A.A.; LastName, B.B.; LastName, C.C. Article Title. *Journal Name* **Year**, *Volume Number*, Page Range.

ISBN 978-3-0365-1604-2 (Hbk)
ISBN 978-3-0365-1603-5 (PDF)

Contents

About the Editors

Chang-Wook Lee

Changwook Lee received his B.S. degree from Kangwon National University, Chuncheon, Korea, and M.S. and Ph.D. degrees from Yonsei University, Seoul, Korea in 1999 and 2009, respectively. He held a postdoctoral position in InSAR for the ARTS contract with the US Geological Survey EROS data center, and all work was performed as determined by the USGS project manager working in coordination with SSSC management. The work was performed at the USGS Cascades Volcano Observatory in Vancouver, Washington, by National Aeronautics and Space Administration (NASA) project supports from 2009 to 2011. He is an associate professor in the department of science of education, Kangwon National University, in Chuncheon, Korea. His research interests include SAR, InSAR and time-series processing technique development for natural disaster monitoring and resource characterization. He has authored more than 100 articles in these research fields.

Hyangsun Han

Hyansun Han received his B.S. and M.S. degrees in geophysics, and Ph.D. degree in remote sensing from the Department of Geophysics, Kangwon National University, South Korea, in 2006, 2008 and 2013, respectively. From March 2013 to September 2014, he was a postdoctoral researcher with the Research Institute for Earth Resource at Kangwon National University. From October 2014 to August 2015, he was a research professor with the Ulsan National Institute of Science and Technology, South Korea. He was a senior research scientist with the Korea Polar Research Institute (KOPRI), South Korea, from September 2015 to February 2020. He is an assistant professor in the Department of Geophysics, Kangwon National University. His research interests include satellite remote sensing of cryosphere, land and ocean, and various applications of remotely sensed data. He received Student Paper Awards at the Korea Remote Sensing Symposia in 2006 and 2007, Excellence Research Awards at Kangwon National University in 2013, and KOPRI's Scientist Awards in 2016, 2017 and 2018.

Hoonyol Lee

Hoonyol Lee received his B.S. degree in geology and M.S. degree in geophysics from the Department of Geological Sciences, Seoul National University, Seoul, South Korea, in 1995 and 1997, respectively, and his Ph.D. degree in radar remote sensing from the Department of Earth Sciences and Engineering, Imperial College London, University of London, London, U.K., in 2001. From 2001 to 2003, he was a postdoctoral research associate with Imperial College London. From 2003 to 2004, he was a senior researcher with the Korea Institute of Geoscience and Mineral Resources (KIGAM), Daejeon, South Korea. He has been with the Department of Geophysics, Kangwon National University since 2004. He was a Visiting Scholar with the Department of Geological Sciences, University of Oregon, Eugene, OR, USA, from August 2008 to July 2009, and with the Department of Civil, Environmental, and Construction Engineering, University of Central Florida, Orlando, FL, USA, from August 2014 to July 2015. He is currently the president of the Korean Society of Remote Sensing since 2021.

Yu-Chul Park

Yu-Chul Park received his Bachelor of Science degree in geology in 1992 from Seoul National University, his master of science degree in geophysics in 1995 from Seoul National University, and his doctor of science degree in mathematical geology from Purdue University in 2000, USA.

Preface to "Artificial Intelligence Methods Applied to Urban Remote Sensing and GIS"

Recently, remote sensing and GIS techniques have gained increasing importance for rapid urbanization, the expansion of urban growth, and the enlargement of populations, due to the application of artificial intelligence, machine learning, and deep learning algorithms. This Special Issue aims to present the state-of-the-art research on optics, SAR, hyperspectral images, and GIS techniques for monitoring urban area environments corresponding to changes in times using publicly available and commercial datasets such as satellite and UAV data.

Given the information above, the aim of this Special Issue is to present the observed urban area and monitor the surrounding urban area in "Artificial Intelligence Methods Applied to Urban Remote Sensing and GIS". This research paper of *Remote Sensing* will cover a wide range of fields, including GISs, remote sensing, earth science, computer science, and environmental science, to analyze the urbanization phenomenon along with theoretical research and practical developments. Some of the prospective/encouraged topics for this Special Issue include:

- Remote sensing applications in urban disaster monitoring using AI;
- Groundwater monitoring in urban areas;
- Fusion of multispectral and SAR image applications;
- Hyperspectral image applications in urban area classification;
- Natural/artificial disaster monitoring;
- Deep/machine learning method algorithms;
- Change detection monitoring in urban areas;
- UAV/drone image processing and analysis;
- Water, river, and lake monitoring in and surrounding urban areas;
- Land subsidence, sink hole, and landslide monitoring;
- Urban river and stream ice monitoring;
- Survey research for citizens'perceptions of urban disaster.

Chang-Wook Lee, Hyangsun Han, Hoonyol Lee, Yu-Chul Park
Editors

remote sensing

Article

Improvement of Earthquake Risk Awareness and Seismic Literacy of Korean Citizens through Earthquake Vulnerability Map from the 2017 Pohang Earthquake, South Korea

Ju Han [1], Arip Syaripudin Nur [2], Mutiara Syifa [3], Minsu Ha [3], Chang-Wook Lee [2,3] and Ki-Young Lee [3,*]

1   Department of Home Economics Education, Kangwon National University, Gangwon-do, Chuncheon-si 24341, Korea; zoz20202@kangwon.ac.kr
2   Department of Smart Regional Innovation, Kangwon National University, Gangwon-do, Chuncheon-si 24341, Korea; aripsyaripudin@kangwon.ac.kr (A.S.N.); cwlee@kangwon.ac.kr (C.-W.L.)
3   Division of Science Education, Kangwon National University, Gangwon-do, Chuncheon-si 24341, Korea; mutiarasyifa@kangwon.ac.kr (M.S.); msha@kangwon.ac.kr (M.H.)
*   Correspondence: leeky@kangwon.ac.kr

Abstract: Earthquake activities in and around the Korean Peninsula are relatively low in number and intensity compared with neighboring countries such as Japan and China. However, recent seismic activity caused great alarm and concern among citizens and government authorities, and uncovered the level of preparedness toward earthquake disasters. A survey has been conducted on 1256 participants to investigate the seismic literacy of Korean citizens, including seismic knowledge, awareness and management using a questionnaire of citizen earthquake literacy (CEL). The results declared that the citizens had low awareness and literacy, which means that they are not properly prepared for earthquake hazards. To develop an earthquake risk reduction plan and program efficiently and effectively, not only must it appropriately characterize the target audience, but also indicate high potential earthquake zones and potential earthquake damage. Therefore, this study mapped and analyzed the seismic vulnerability in southeast Korea using LogitBoost, logistic model tree (LMT), and logistic regression (LR) machine learning algorithms based on a building damage inventory map. The damaged buildings' locations were generated after the 2017 Pohang earthquake using the damage proxy map (DPM) method from the Sentinel-1 synthetic aperture radar (SAR) data. DPMs detected coherence loss, which indicates damaged buildings in urban areas in the Pohang earthquake and shows a good correlation with the Korea Meteorological Administration (KMA) report with modified Mercalli intensity (MMI) scale values of more than VII (seven). The damage locations were randomly divided into two datasets: 50% for training the vulnerability models and 50% for validating the models in terms of accuracy and reliability. Fifteen seismic-related factors were used to construct a model of each algorithm. Model validation based on the area under the receiver operating curve (AUC) was used to determine model accuracy. The AUC values of seismic vulnerability maps using the LogitBoost, LMT, and LR algorithms were 0.769, 0.851, and 0.749, respectively. We suggest that earthquake preparedness efforts should focus on reconstruction, retrofitting, renovation, and seismic education in areas with high seismic vulnerability in South Korea. The results of this study are expected to be beneficial for engineers and policymakers aiming at developing disaster risk reduction plans, policies, and programs due to future seismic activity in South Korea.

Keywords: seismic vulnerability map; DPM method; Sentinel-1; machine learning; seismic literacy

Citation: Han, J.; Nur, A.S.; Syifa, M.; Ha, M.; Lee, C.-W.; Lee, K.-Y. Improvement of Earthquake Risk Awareness and Seismic Literacy of Korean Citizens through Earthquake Vulnerability Map from the 2017 Pohang Earthquake, South Korea. *Remote Sens.* 2021, *13*, 1365. https://doi.org/10.3390/rs13071365

Academic Editor: Peter Hofmann

Received: 4 March 2021
Accepted: 27 March 2021
Published: 2 April 2021

Publisher's Note: MDPI stays neutral with regard to jurisdictional claims in published maps and institutional affiliations.

1. Introduction

Earthquake activities in and around the Korean Peninsula are relatively low in number and intensity compared with neighboring countries such as Japan and China, because it is located within the Eurasian intracontinental region [1]. However, seismographs often record sudden occurrences of moderate earthquakes; historical documents show that several damaging earthquakes happened in the country [2], indicating that the Korean Peninsula is not completely safe from earthquake disasters.

On 15 November 2017, an M_L 5.4 earthquake occurred in Pohang, South Korea at 05:29:31 UTC [3], causing widespread damage in and around the city [4]. The earthquake was the second largest to occur in the Korean Peninsula since earthquake monitoring was initiated by the Korea Meteorological Administration (KMA) in 1978 [4,5]. In terms of the magnitude, the Pohang earthquake was not larger than the Gyeongju earthquake. However, the damage of the Pohang earthquake was much more than that of the Gyeongju earthquake. The Pohang earthquake caused more than USD 75 M of indirect damage to over 57,000 structures and over USD 300 M of total economic impact, as estimated by the Bank of Korea, injured 135 residents, and displaced more than 1700 people into emergency housing [6]. Meanwhile, the Gyeongju earthquake resulted in approximately USD 9.5 M in damage to 5368 properties, and 23 injured people [7]. More than 100 heritage buildings and monuments sustained damage from the earthquakes [8]. Twenty-one kilometers from Pohang, there is the Gyeongju Historic Area that was registered as a UNESCO World Cultural Heritage Site in November 2000, an area that embodies the time-honored history and culture of Gyeongju, the ancient capital of the Silla Kingdom (57 BC–935 AD). Some damage was found in this area, such as to Dabotap Pagoda (dislocated banister), Cheomseongdae Observatory (shifted and tilted), and Gyeongju Gyochon Traditional Village (cracked walls) [9]. Therefore, we need to conduct research about it.

Moreover, satellite images could detect surface deformation after the earthquake in Pohang, and therefore, for the first time, surface deformation measurements for an earthquake in Korea historically. A radar interferometry image was taken by the Sentinel-1 satellite, highlighting a deformation of −5 to 5 cm (blue to red) that occurred near Pohang city center. This image was obtained from a two-pass interferogram created using GAMMA software [10]; in this process, two synthetic aperture radar (SAR) images (4 and 16 November 2017) were co-registered to form an interferometric pair, which were then cropped to the area of interest. Topographic interferometric synthetic aperture radar (InSAR) images were then produced through interferogram generation. These images were derived from global 1 arcsecond Shuttle Radar Topography Mission (SRTM) data (30 m resolution), followed by topographic phase removal and differential synthetic aperture radar interferometry (DInSAR) phase generation, leaving only the deformation phase image [11]. A phase unwrapping procedure was then applied to generate an unwrapped DInSAR image, for conversion into displacement values (cm). Considering the uniqueness of this earthquake, we choose Pohang earthquake as our study case.

Additionally, we also need to know citizens' risk awareness about earthquakes in Korea as they had never experienced or felt the degree of natural hazard caused by the Pohang earthquake directly before. However, many people were not aware of earthquakes risk from the survey data. A survey was conducted with 1256 Korean citizens during spring 2020. Figure 1 shows the survey results that only 6% of the participants were aware that where their live is "absolutely not at all" safe from earthquakes, 29% of the participants thought "hardly not", 42% thought "normal", 20% thought "to some extent", and 3% thought they were "absolutely" safe from earthquakes. These results are an important warning sign for regulators and authorities, given the recent earthquakes that caused great human and material losses.

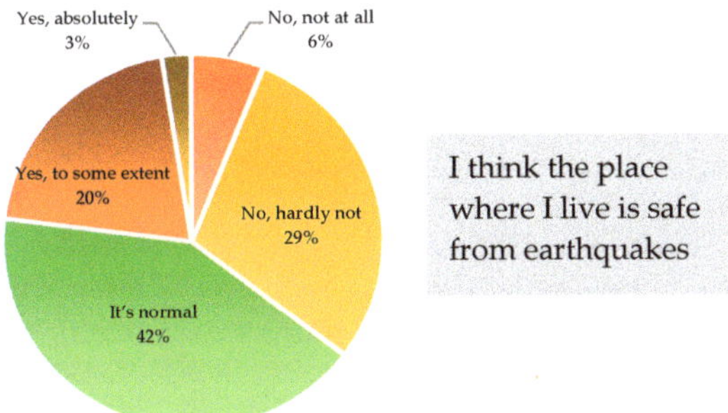

Figure 1. Chart of earthquake risk awareness of Korean citizens.

However, a small number of people in the earthquake area (close distance) were aware of the extreme fear and danger of earthquakes compared to the majority of people according to the survey results. Figure 2 shows that people who live near the epicenter of the Pohang earthquake have higher awareness of earthquakes because they were directly affected by the damage from the Pohang earthquake. Meanwhile, people who live far from the epicenter, such as people living in Seoul, have a lower awareness level of earthquakes because they only felt a slight shock from the Pohang earthquake, which was not fatal. Therefore, we need to inform many people of the dangers caused by such an earthquake, and when earthquakes occur in another area in the future, we need to be able to recognize the earthquake and respond to it.

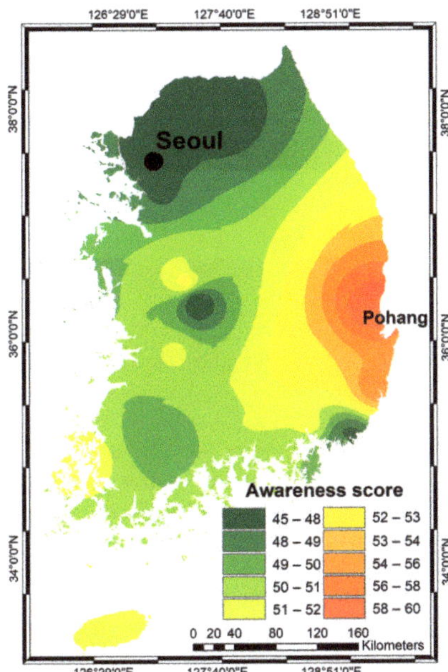

Figure 2. Map of citizens' awareness risk awareness score throughout Korea.

To reduce the effects of earthquake disasters and develop an earthquake risk reduction plan and program efficiently and effectively, not only is performing sustainable preparation, such as seismic literacy of citizens, necessary, but so is indicating high potential earthquake zones and potential earthquake damage by producing seismic vulnerability maps. Seismic vulnerability assessment involves the comprehensive evaluation of factors that affect risks associated with earthquakes within predefined areas. Urban areas are at higher risk of seismic disasters than outlying areas due to their higher building and infrastructure density and larger population. Therefore, in assessing seismic vulnerability, it is essential to select suitable influential factors and methods for the area of interest. Several methodologies have been applied for seismic vulnerability assessment and mapping during the past few decades [7,12–15].

Seismic vulnerability assessment studies commonly analyze case studies using a combination of multicriteria decision making (MCDM) and geographic information system (GIS) approaches [12,13,16,17]. Among these, the analytical hierarchy process (AHP) is one of the most widely known MCDM methodologies; it stratifies and quantifies the importance of each applied influential factor to determine its relative importance, and assesses vulnerability by applying weights to all factors [12,16,18]. However, this method can be subjective because the opinion of the researcher can affect the weight assignment process; therefore, it is somewhat unsuitable for objective assessment. To address this problem, recent studies have applied hybrid models that combine various methodologies [14,19].

Many recent studies related to seismic vulnerability assessment and mapping have been conducted using machine learning techniques [12,16,18–21]. For example, Han et al. (2019) [20] used a logistic regression (LR) model and applied the support vector machine (SVM) methodology to four kernel models (linear, polynomial, radial basis function, and sigmoid) to derive a suitable model for seismic vulnerability assessment; this study was notable in that the results of several seismic vulnerability models were compared analytically; such analyses are rarely conducted in this field, despite the broad application of machine learning techniques in recent years.

Providing training data plays an important role in the accuracy of the vulnerability map. Here, we used the damage proxy map (DPM) method to extract a building damage map for training and testing datasets. The DPM method is part of an ongoing collaborative effort between the Jet Propulsion Laboratory (JPL) and the California Institute of Technology, called the Advanced Rapid Imaging and Analysis (ARIA) project. The DPM method, using synthetic aperture radar (SAR) satellite data, has been shown to be useful for damage mapping following an earthquake and other natural disaster events, including the 2015 MW 7.8 Gorkha, Nepal earthquake using COSMO-SkyMed and ALOS-2 satellites [20], the 2019 typhoon Hagibis, Japan using Sentinel-1 satellites [21], the 2019 MW 7.1 Ridgecrest earthquake in California using Sentinel-1 satellites [22], and the 2014 eruption of Kelud volcano (Indonesia) using COSMO SkyMED satellites [23].

This study aims to improve the risk awareness and seismic literacy of Korean citizens through an earthquake vulnerability map of all buildings in southeast Korea. To produce the earthquake vulnerability map, we generated a damage proxy map (DPM) after the 2017 Pohang earthquake from the Sentinel-1 SAR dataset as a dependent variable, then applied machine learning to construct models using 15 seismic-related factors as independent variables. Model performances were verified using a receiver operating characteristic (ROC) curve. Finally, dangerous and safe areas were identified in southeast Korea by creating maps based on the model with the highest accuracy for each methodology, and the result were assessed. The results of this study should improve citizen earthquake risk awareness and seismic literacy, especially in high seismic vulnerability areas, and facilitate the construction of seismic vulnerability models that will be useful to reduce future losses due to earthquakes in South Korea.

2. Materials and Methods

2.1. Study Area

The study area covers three metropolitan cities; Busan, Daegu, and Ulsan, and two provinces; Geyongsangbuk and Gyeongsangnam, surrounding the epicenter of the Pohang earthquake. For simplicity, we refer to it as southeast Korea. The blue line in Figure 3 shows the border of the study area. In total, southeast Korea is home to 12,961,687 people and has an area of 32,285 km^2 [24]. Within the total area, urban areas account for 5.49%, followed by agriculture at 21.88%, forestry (66.71%), and other areas (5.92%). The proportions of males and females in these areas are 50.21% and 49.79%, respectively [25].

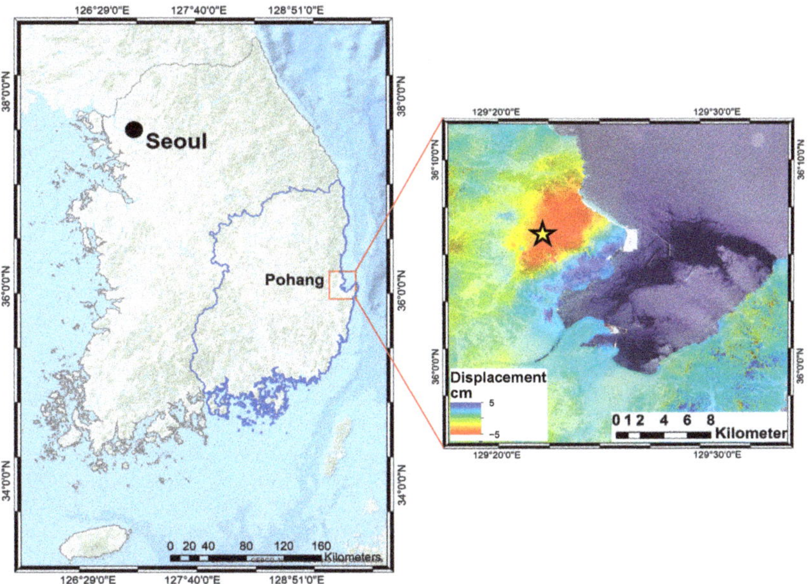

Figure 3. Study area of this study, indicated by a blue line. Location of the 2017 Pohang earthquake and surface deformation generated from Sentinel-1 synthetic aperture radar (SAR) data acquired on 4 and 16 November 2017.

The study area was chosen considering the high seismic activity in these areas. A series of aftershocks of the Pohang earthquake was observed to have occurred with a magnitude of 3.0 or more until 31 May 2018. A magnitude of 4.3 occurred near the epicenter, 2 h after the mainshock, and a magnitude 4.6 earthquake occurred 4 km to the southwest at around 20:03:04 UTC on 11 February 2018 [26]. One year prior to the Pohang earthquake, an M_L 5.8 earthquake shocked Gyeongju at 11:32:55 UTC on 12 September 2016. The epicenter was 8.7 km from the south of the city and 15 km beneath the surface [27]. The Gyeongju earthquake was just 40 km from the site of the Pohang earthquake. The Gyeongju earthquake was accompanied by 600 aftershocks, including an M_L 5.1 foreshock that occurred near the mainshock at 10:44:32 UTC [28]. In 2018, 115 earthquakes with magnitudes of more than 2 occurred in the Korean Peninsula; among these, 36 earthquakes (31.13%) occurred in southeast Korea [29]. In 2019, 957 earthquakes with a magnitude of less than 2.0 occurred in the Korean Peninsula; among these, 294 earthquakes (30.72%) occurred in southeast Korea [30]. Among the 88 earthquakes of magnitude 2.0 or higher, 23 (26.17%) occurred in the same area.

The Korean Peninsula lies at the eastern margin of the Eurasian Plate. About 30–15 million years ago, north-northeast (NNE)-striking strike–slip faults and NNE- to NE-striking normal faults settled predominantly in southeastern Korea and adjacent offshore areas when the East Sea opened in the early to middle Tertiary as a back-arc basin; smaller-scale coetaneous basins also formed, including the Pohang Basin. Although this region is about 400–500 km in length, its seismicity is affected by complex interactions of the Indo-Australian and Eurasian plates, as well as by subduction of Philippine Sea plates beneath the Japan and Ryuku trenches [31]. Several faults are distributed within the study area, including Dongrae, Moryang, Miryang, Ulsan, Wangsan, and Yangsan [32]. Due to seismic history and geographic characteristics, the probability of earthquake occurrence in southeast Korea is considered relatively high, and secondary damage in the event of an earthquake with a medium or higher magnitude constitutes an unusually high risk. Therefore, sustainable preparation and management planning for such events is required.

2.2. SAR Datasets

A building damage inventory map for producing the seismic vulnerability map in southeast Korea was generated using Sentinel-1 synthetic aperture radar (SAR) C-band data (5.5 cm wavelength) provided by the European Space Agency (ESA). Pohang is located in between two frames (470 and 475), path 61 of the Sentinel-1 imagery; therefore, we obtained six Sentinel-1 single look complex (SLC) images with vertical transmission and vertical return (VV) polarization for the Pohang earthquake, with four scenes prior to the event on 23 October 2017, and 4 November 2017, and two scenes after the event on 16 November 2017. Furthermore, we needed to merge the scenes before processing. The images were co-registered, with the 4 November 2017 scenes as a reference.

2.3. Damage Proxy Map (DPM)

DPMs generated from the comparison of pre- and co-event SAR images can help identify damage caused by earthquakes using remote sensing imagery [20]. The method relies on the reduction in the coherence of the radar echoes between satellite-based SAR images taken before and after the earthquake to identify anomalous changes in ground surface properties. Coherence measures the change in radar backscatter from the ground, a proxy for the ground-surface property changes. A low coherence implies a large change to the ground surface that reflected the SAR radiation [33]. Changes can be caused by damage to the ground itself or damage to structures.

The process started with image co-registration, with the 4 November 2017 scenes as a reference. This co-registration process was done with sub-pixel accuracy to match scenes to one another. We used the complex pixel value, c, of the pre-processed SLCs for the change detection analysis, where damage is inferred from loss of coherence or decorrelation between SAR images [20]. We computed the pre—and co-event interferometric coherences, γ (Equation (1)), from a pair of SLCs before the event and another pair spanning across the event, respectively [21].

$$\gamma = \frac{|\langle c_1 c_2^* \rangle|}{\sqrt{\langle c_1 c_1^* \rangle \langle c_1 c_2^* \rangle}}, 0 \leq \gamma \leq 1 \tag{1}$$

where c_1 and c_2 are complex pixel values of two co-registered SAR images and * denotes the complex conjugate. The resulting coherence ranges from 0 (incoherent) to 1 (coherent). The coherence is equal to 1 if the observation is identical in the two images because of the stable object-like buildings in the scene. The pre-event coherence represented change unrelated to the event and was assumed to be the background value. Then, we obtained a coherence difference (COD) by subtracting $\gamma_{\text{pre-event}}$ from $\gamma_{\text{co-event}}$. Therefore, the process could generate a COD ranging from -1 to 1. A negative COD (or coherence gain) usually indicates surface changes occurring between the pre-event scenes and is associated with changes not related to the event. Coherence gain could happen in an agricultural area when the fields are full of crops. Then, when harvested during the period time of the pre-event interferometric pair, the area has low coherence. After harvesting, leaving an empty field, the area has a greater coherence in the co-event interferometric pair ($\gamma_{\text{co-event}} > \gamma_{\text{pre-event}}$), so a negative COD is obtained. A positive COD (or coherence loss) indicates surface changes between the co-event scenes spanning the events, such as major damage to a building significantly that increases the interferometric phase variance, causing a decrease in coherence. Hence, the loss of coherence is most effective for detecting damage in built-up areas caused by earthquake. However, COD is generally less effective and less reliable in vegetated areas where coherence changes may be random. Therefore, this study focused on a DPM in urban, built-up areas as changes can be detected easily in SAR imagery. A greater loss in coherence generally correlates with greater severity of the change, such as a fully collapsed building, causing more significant coherence loss than partial collapse [21].

The threshold for significant coherence loss can be chosen by comparing observed coherence changes with reported damage and areas in which it is known that no damage occurred. Yun et al. (2015) compared a DPM with a National Geospatial-Intelligence Agency (NGA) analysis and the United Nations Operational Satellite Application Programme (UNOSAT) damage assessment map. Tay et al. (2020) used high-resolution aerial imagery from the Geospatial Information Authority of Japan (GSI). Here, coherence loss thresholds for DPMs were chosen by considering the damaged area from the Korea Meteorological Administration (KMA) report of the Pohang earthquake [34].

2.4. Selection of Seismic-Related Factors

We selected factors affecting seismic vulnerability based on the result of previous studies [7,14,15,17]. The factors affecting the seismic vulnerability were prepared based on five main indicators, which were geotechnical, physical, structural, social, and capacity; we selected a total of 15 factors corresponding to these categories. Geotechnical factors included slope and altitude; physical factors included peak ground acceleration (PGA), epicenter distance, and fault distance; structural factors included land use, construction materials, building density, and building height; social factors included elderly population ($\geq$65 years), child population (<15 years), and population density; and capacity factors included distances from hospitals, fire stations, and police stations. The factors were organized into raster-based spatial databases (30 m spatial resolution) and were reclassified using the quantile method to identify and analyze the effect of each class. The data used in this present study are shown in Figure 4.

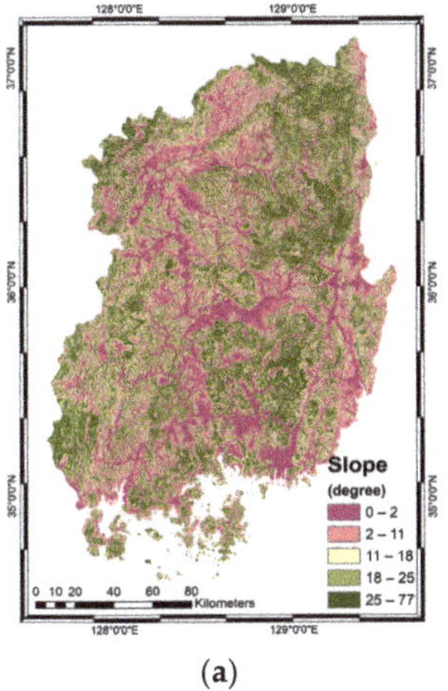

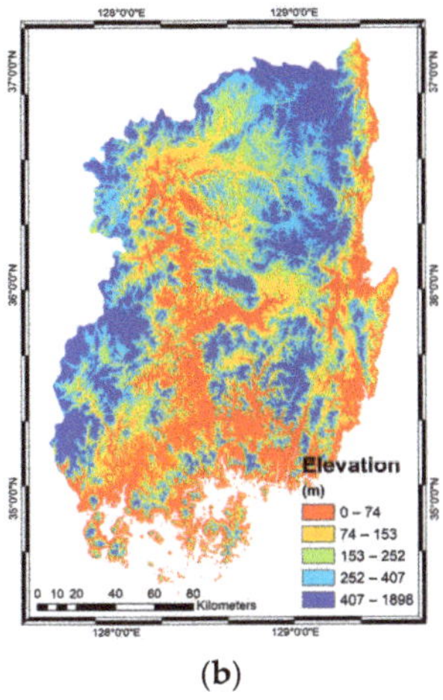

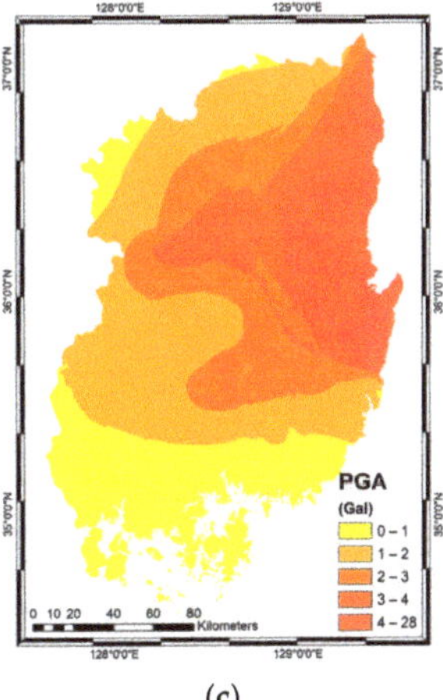

(a)   (b)   (c)

Figure 4. *Cont.*

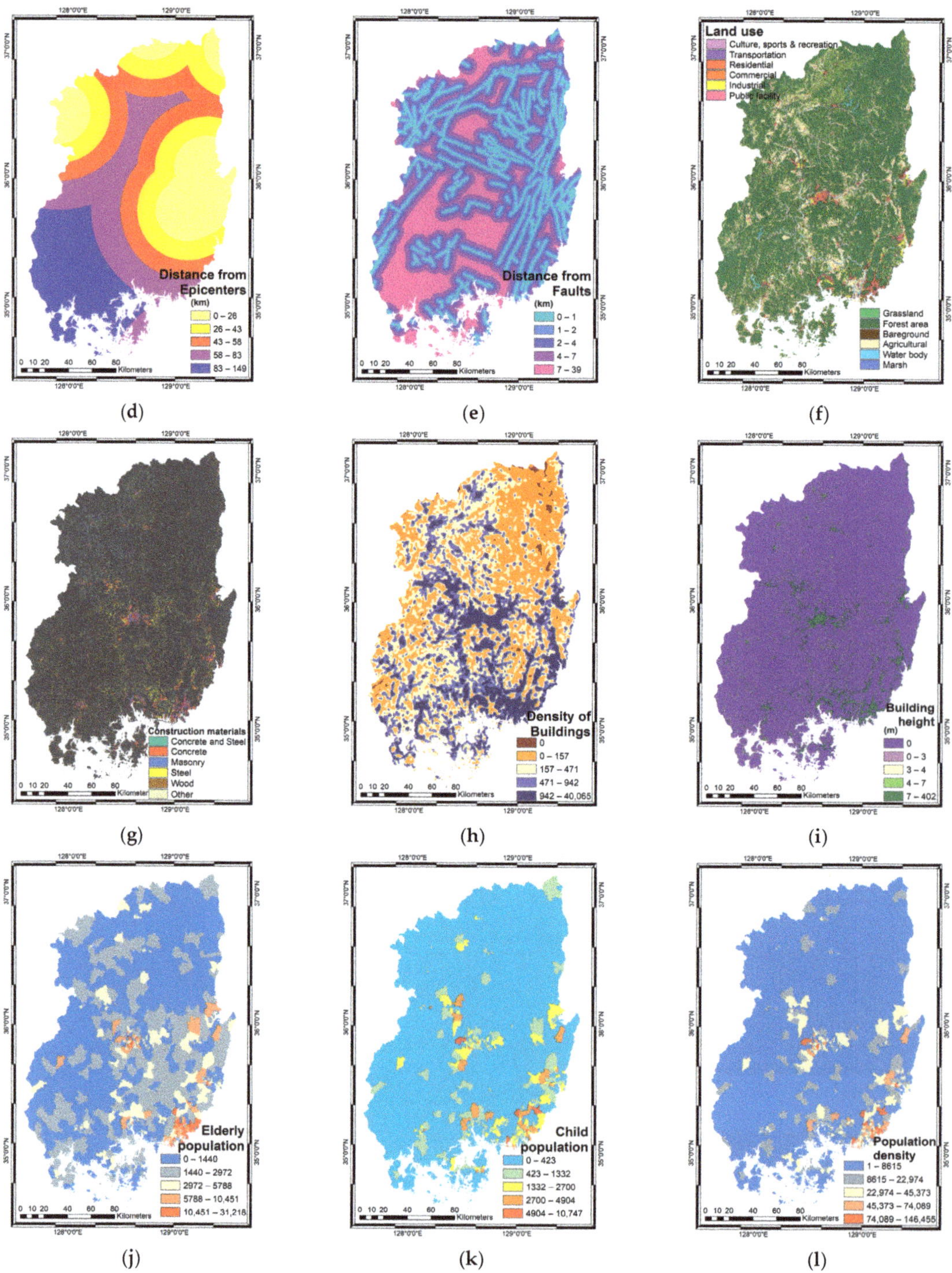

(**d**) (**e**) (**f**) (**g**) (**h**) (**i**) (**j**) (**k**) (**l**)

Figure 4. *Cont.*

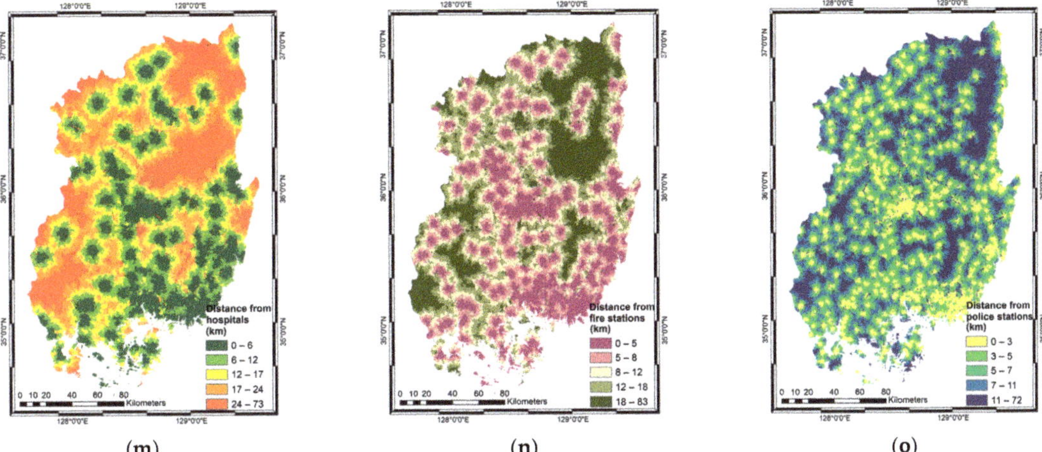

(m)   (n)   (o)

Figure 4. Factors related to seismic vulnerability: (**a**) slope, (**b**) elevation, (**c**) peak ground acceleration (PGA), (**d**) distance from epicenters, (**e**) distance from faults, (**f**) land use, (**g**) construction materials, (**h**) density of buildings, (**i**) building height, (**j**) elderly population, (**k**) child population, (**l**) population density, (**m**) distance from hospitals, (**n**) distance from fire stations, and (**o**) distance from police stations.

Slope and elevation were extracted from a digital elevation model (DEM) of the Shuttle Radar Topography Mission (SRTM) using basic terrain analysis tools. Slope and elevation are factors affecting the vulnerability of urban environments to earthquakes [14,35]. Degradation in terrain with steep topography, especially at the top of hills and peaks, is greatly enhanced. According to construction standards, a slope of 5 to 9% is suitable for urbanization [16]. The highs and lows of elevation in each area are highly correlated with landslide susceptibility in each area [36]. Therefore, because of the amount of erosion and its relation to human activity, the higher the altitude of an area, the greater the seismic vulnerability. Figure 4a,b shows the slope and elevation maps of southeast Korea, respectively.

Peak ground acceleration (PGA) (Figure 4c) is the degree to which the ground shakes at the Earth's surface and is related to the amount of fault activity [18]. In this study, raw data from the Korea Institute of Geoscience and Mineral Resources (KIGAM) were converted to acceleration data and interpolated throughout southeast Korea [27,37]. The epicenter, or location where an earthquake occurs, is an important factor related to earthquake occurrence; the level of damage is different depending on the ground condition or the structure of the fault plane, on which the greatest damage often occurs at the epicenter. Therefore, we used distance data from earthquake epicenters (Figure 4d) with a magnitude over 4 from 1978 to 2020, including the 2016 Gyeongju earthquake and the 2017 Pohang earthquake. The epicenter locations were acquired from the United States Geological Survey (USGS). The faults are forms of tectonic factors whose presence or absence can be examined in relation to the seismic hazard of different areas. Fault distance (Figure 4e) plays a key role in vulnerability to earthquake hazards, as proximity to the fault causes high seismic risk and damage, and distance from it will reduce the risk and consequently provide higher resilience [16].

Anti-seismic design in South Korea was introduced in 1988, and it is mandatory only for buildings that are three stories or higher [38]. As of November 2016, 29.9% of residential buildings and 23.7% of non-residential buildings in Seoul were designed to be anti-seismic. As there is no guarantee that future earthquakes will not exceed the magnitude of the Gyeongju earthquake, most buildings in South Korea are considered highly vulnerable. To assess their vulnerability, we identified four structural factors of seismic vulnerability: land use, construction materials, building density, and building height, depicted in Figure 4f–i, respectively [7,17]. The greater the number of floors of a building, despite its quality, the greater the vulnerability. The number of floors in the building, if not in accordance with safety principles, will definitely increase the damage [39]. Even if the height is treated with due diligence and calculations, it is difficult for the evacuation of buildings, and to search for and rescue people. Obviously, structures of high strength and standard materials have good earthquake safety [40]. Proper deployment of land uses on the basis of urban planning principles, such as proper accessibility, proper distance from the biological hotspots, safety, comfort, and utility can substantially reduce the amount of vulnerability, injury, and economic damage [41].

Increasing population growth, population density, and poor distribution of services and infrastructure pose risks to society [42]. In recent earthquakes around the world, it can be said that most of the damage is to humans and with the increase in population, it is predicted that in the future, the mortality rate will be higher. The earthquake hazard coefficient in urban centers is also more complex and riskier due to urbanization without planning and development [43]. In events such as earthquakes, everyone in the community is vulnerable, but older people and children are the most vulnerable groups in a community and more attention is needed to minimize pain and injury [44]. Children do not tolerate disruption well and older people are psychologically fragile because of their disrupted life rhythms. The elderly population, child population, and population density in southeast Korea are presented in Figure 4j–l, respectively.

We identified the locations of social infrastructure facilities that offer aid in the event of an earthquake, and of hazardous facilities that have the potential to cause huge damage. The degree of accessibility following a disaster was analyzed by considering the physical distances to three factors, including social infrastructure facilities (hospital, police station, and fire station). Distance from a hospital (Figure 4m) and access to health services such as hospitals play a key role in controlling post-emergency complications and providing earthquake rescue and hospitalization services. Proper and quick access to medical facilities will increase earthquake resilience [45]. Distance from a fire station (Figure 4n) and access to police stations (Figure 4o) through the communication networks will speed up rescue operations and service the injured. As such, the greater the distance from fire stations and police stations, the greater the vulnerability [46].

2.5. Machine Learning

To map seismic vulnerability using a machine learning algorithm, several steps must be performed. First, the spatial relationships between the damaged buildings from the DPM and related factors (geotechnical, physical, structural, social, and capacity indicators) are calculated using the frequency ratio (FR) method. In this research, we analyzed the spatial relationship between damaged buildings' locations (1623 cells) and 15 factors related to seismic vulnerability, based on the FR value of each factor. When the ratio is greater than 1, this denotes that the class in each factor has a closer relationship with seismic vulnerability [47]. In the case that each factor has a less close relationship with seismic vulnerability, the ratio is less than 1. The FR for each factor was calculated by Equation (2) [48].

$$FR = \frac{\% \text{ of class of related factor}}{\% \text{ of total area}} \tag{2}$$

To apply the machine learning algorithm, we used the 1623 cells of damaged buildings generated from the DPM method. Among these cells, 50% (812) were used as a training dataset and 50% (811) were used as a test dataset. We extracted the same number of cells corresponding to undamaged buildings. All cells were randomly sampled and generating models and the accuracy of each model was done based on training (1624) and test datasets (1622). Several seismic-related maps, including geological maps, were produced at a 30 m resolution. Then, all data were classified as categorical or continuous. Continuous variables include the slope, elevation, PGA, distance from epicenters, distance from faults, building density, building height, child population, elderly population, population density, distance from hospitals, distance from police stations, and distance from fire stations. Categorical variables include construction materials and land use.

Model validation was carried out through ROC curve analysis of the testing dataset (50%). Receiver operating characteristic (ROC) curve analysis, as an index of model performance, is commonly used to assess predictive accuracy [49]. To quantitatively determine the accuracy of the model verification, the area under the curve (AUC) of the ROC curve is calculated for the total area and correct predictive accuracy is obtained. AUC values range between 0.5 and 1; higher values indicate more reliable algorithm performance. The workflow of the seismic vulnerability mapping carried out in this study is provided in Figure 5.

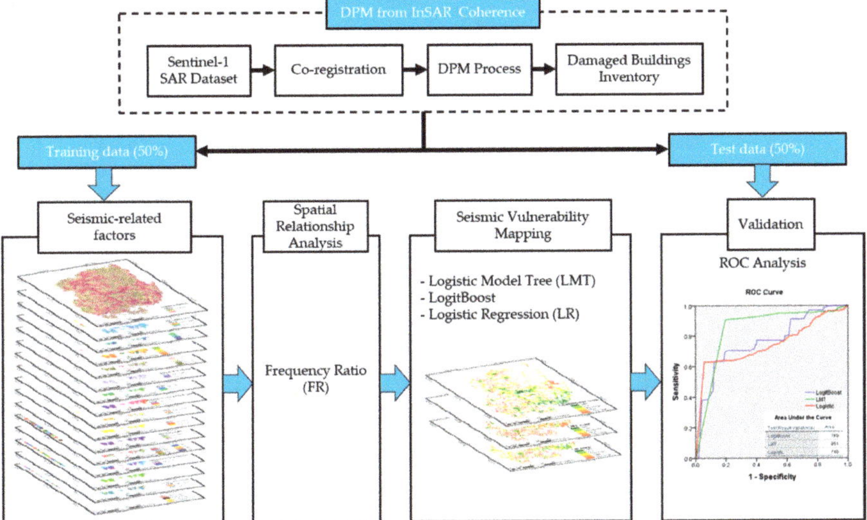

Figure 5. Workflow of seismic vulnerability mapping. Abbreviations: DPM, damage proxy map; InSAR, interferometric synthetic aperture radar; ROC, receiver operating characteristic.

2.5.1. LogitBoost

LogitBoost is a boosting algorithm developed by Friedman et al. [50] to reduce bias and variance. The LogitBoost algorithm was modified from AdaBoost, which was the commonly used boosting method for handling noisy data that executes additive logistic regression with least-square fits for individual classes. LogitBoost reduces training errors and enhances classification accuracy by using additive logistic regression for classification with a base-learning regression scheme and an ability to perform multiclass classification. The damaged building inventory map was divided into two classes: damaged buildings and undamaged buildings, using Equation (3):

$$Lc(c) = \sum_{i=1}^{D} \beta_i x_i + \beta_0 \tag{3}$$

where D is the number of building damage-dependent factors and β_i is the coefficient of the i-th component within input vector x. Probabilities were constructed using the linear logistic regression method with Equation (4):

$$P\left(\frac{C}{x}\right) = \exp(Lc(x)) / \sum_{C'=1}^{C} \exp(Lc'(x)) \tag{4}$$

where C is the number of classes and the least-square fit Lc(x) is resolved such that $\sum_{C=1} L_C^C(x) = 0$ to set up the least number of instances per node of the logistic model trees.

2.5.2. Logistic Model Tree (LMT)

The logistic model tree combines the C4.5 algorithm [51] and logistic regression (LR) functions. The information gain ratio (IGR) technique is applied to split the tree into nodes and leaves, and the LogitBoost algorithm [52] is used to fit the logistic regression functions at a tree node. The C4.5 algorithm uses the entropy technique for feature selection because it is the fastest method for providing reliable classification accuracy [53]. The over-fitting problem, which is an important challenge in LMT modeling, is overcome using the CART algorithm, which prunes the tree for modeling the training dataset [54]. The IGR can be formulated using Equation (5):

$$\text{Gain ratio (A)} = \frac{\text{gain (A)}}{\text{split info (A)}} \tag{5}$$

where gain (A) is the information after attribute A is selected as a test for classification of the training samples and split info (A) is the information generated when x training samples are categorized into n subsets [51]. In the next step, the LogitBoost algorithm performs additive logistic regression with least-squares fit for each class Ci (damaged or undamaged building) according to Equation (6) [55]:

$$L_c(x) = \sum_{i=1}^{CF} \alpha_i x_i + \alpha_0 \tag{6}$$

where $L_c(x)$ is the least-squares fit and CF and α_i are, respectively, the number of seismic-related factors and the coefficient of the i-th element of vector x. The a posteriori probabilities in the leaves of the LMT are calculated using the linear logistic regression model with Equation (7) [52]:

$$p(c|x) = \frac{\exp(L_c(x))}{\sum_{c'}^{c} \exp(L_{c'}(x))} \tag{7}$$

where c is the number of building damage classes and Lc(x), the least-squares fit, is transformed in such a way that $\sum_{c'=1}^{c} L_c(x) = 0$.

2.5.3. Logistic Regression (LR)

The logistic regression (LR) model, developed by McFadden (1973) [56], is a multivariate regression analysis model that describes the relationship between a bivariate dependent parameter and several independent parameters [57] through the estimation of an optimal model. The addition of a link function suitable for a general linear regression model allows the parameter type to be continuous, discrete, or mixed, thus obviating the requirement for a normal distribution [58,59]. Some studies have shown that the LR model is more accurate than other types of models constructed for the same purpose [60–62]. The LR model based on a general linear model can be derived from Equations (8) and (9):

$$y = b_0 + b_1 x_1 + b_2 x_2 + b_3 x_3 + \cdots + b_n x_n \tag{8}$$

$$P = \frac{e^y}{1 + e^y} \tag{9}$$

where y is the linear logistic model, b_0 is the y-intercept, b_n is the logistic coefficient of each factor, n is the number of factors controlling a seismic event, x is the earthquake conditioning factor, and P is the probability of damage (ranging from 0 to 1) in the event of an earthquake [60].

3. Results

3.1. Building Damage Inventory Map

The DPMs were generated from the Sentinel-1 dataset and geocoded to the Shuttle Radar Topography Mission (SRTM) digital elevation model (DEM), 1 arcsecond. Figure 6 reveals the map view of Sentinel-1 DPMs over Pohang. The assessment technique is most sensitive to the destruction of the built environment. Pixels are set to be relatively transparent where corresponding to areas where decorrelation did not significantly change during the time spanning the earthquake, suggesting little to no destruction. Increased opacity of the radar image pixels reflects increasing ground and building change or potential damage. The color range from yellow to red indicates an increasingly significant coherence change in the area covered by the pixel. Each pixel in the DPMs was registered to the SRTM DEM and had a corresponding dimension of about 30 m.

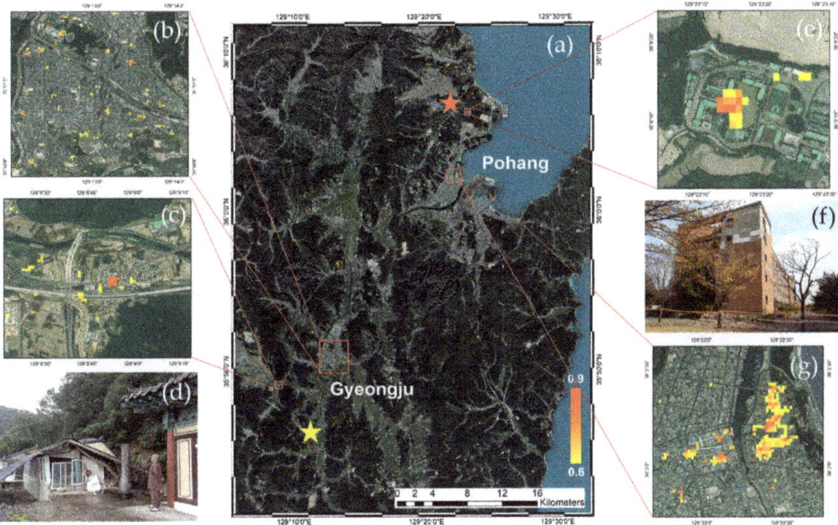

Figure 6. (**a**) Damage proxy maps (DPMs) after the Gyeongju and the Pohang earthquakes. DPM of (**b**) Seonggeon-dong and (**c**) Gwangmyeong-dong corresponding to (**d**) collapsed houses. DPM of (**e**) Handong Global University corresponding to (**f**) collapsed walls and in (**g**) Songdo-dong. Yellow to red pixels indicate increasingly more significant potential damage in seismic vulnerability mapping. Red and yellow stars indicate the epicenter of Pohang and Gyeongju earthquakes, respectively.

Here, we compared the distribution of coherence loss areas after the earthquakes throughout Gyeongju and Pohang, as seen in Figure 6a. According to the land cover map derived from the Korea Institute of Geoscience and Mineral Resources (KIGAM), Figure 6a reveals the DPM after the Gyeongju earthquake over residential, commercial, agricultural, and vegetated areas. Figure 6b,c show the widespread COD in Seonggeon-dong and Gwangmyeong-dong, respectively, corresponding to collapsed houses (Figure 6d [63]). These areas consist of residential and commercial areas. Figure 6a shows DPM that indicates COD over residential, commercial, industrial, agricultural, and vegetated areas affected by the Pohang earthquake. Some of the spatially large and strong coherence loss after the Pohang earthquake corresponds to Handong Global University (Figure 6e), associated with collapsed walls (Figure 6f), while Figure 6g shows Songdo-dong district, consisting of residential areas. The DPM of the Pohang earthquake was then used as a building damage inventory map to produce a seismic vulnerability map using machine learning.

3.2. Relationship between Damaged Buildings and Related Factors

FR values can provide information on the relationship between seismic vulnerability and related factors (geotechnical, physical, structural, social, and capacity). A ratio greater than 1 denotes that the class in the related factor has more impact on seismic vulnerability. The FR values calculated in this study are shown in Table 1. Building damage occurred in areas with elevations of 0–74 m. The classes with a strong impact on seismic vulnerability were: slope of 0°–2.44° (FR = 3.50), elevation of 0–74 m (4.09), PGA of 3.37–28.56 gal (5.31), 0–26.90 km distance from epicenter (5.12), 7.36–39.08 km distance from fault (2.65), building constructed from concrete (2.63), building density of 1256–40,065 (5.13), building height of 7.88–422 m (2.31), commercial areas (3.99), child population of 228–10,747 (3.65), elderly population of 2252–31,218 (3.56), population density of 10,913–146,455 (5.17), 0–6.92 km distance from hospital (3.95), 0–3.39 km distance from police station (3.95), and 0–5.24 km distance from fire station (4.18). Areas in these classes are predicted to experience the highest degree of damage due to earthquakes.

Table 1. Frequency ratios of seismic-related factors.

Factor	Class	Total %	Event %	Frequency Ratio
Slope (degree)	0–2.44	20.06	70.26	3.50
	2.44–11.60	20.21	24.47	1.21
	11.60–18.32	20.14	4.21	0.21
	18.32–25.04	19.79	0.92	0.05
	25.04–77.89	19.77	0.13	0.01
Elevation (m)	0–74	20.02	81.79	4.09
	74–153	20.04	13.7	0.69
	153–252	19.99	4.48	0.23
	252–407	19.97	0	0
	407–1898	19.96	0	0
PGA (gal)	0–0.81	17.97	0	0
	0.81–1.48	20.71	0	0
	1.48–2.26	23.86	0	0
	2.26–3.37	19.26	3.58	0.19
	3.37–28.56	18.17	96.41	5.31
Distance from epicenter (km)	0–26.90	19.55	100	5.12
	26.90–43.27	19.96	0	0
	43.27–58.48	19.88	0	0
	58.48–83.04	20.61	0	0
	83.04–149.12	19.97	0	0
Distance from fault (km)	0–0.92	20.00	18.58	0.93
	0.92–2.29	20.00	8.33	0.42
	2.29–4.14	20.00	6.41	0.33
	4.14–7.36	19.99	13.71	0.69
	7.36–39.08	19.99	52.94	2.65
Construction materials	Steel	31.88	22.81	0.71
	Masonry	28.81	20.70	0.71
	Concrete	21.08	55.57	2.63
	Wood	17.86	0.9	0.05
	Concrete and steel	0.25	0	0
	Other	0.10	0	0
Building density	0–157	11.85	0	0
	157–314	29.08	0.51	0.02
	314–628	22.51	7.56	0.34
	628–1256	21.96	17.30	0.79
	1256–40,065	14.57	74.61	5.13
Building height (m)	0–1.57	20.11	3.55	0.18
	1.57–3.15	20.76	7.32	0.36
	3.15–4.73	19.70	14.43	0.74
	4.73–7.88	19.72	29.28	1.49
	7.88–402	19.68	45.39	2.31

Table 1. *Cont.*

Factor	Class	Total %	Event %	Frequency Ratio
Land use	Residential	2.70	37.64	0.76
	Industrial	0.81	13.64	0.92
	Commercial	0.32	23.29	3.99
	Culture, sports, and recreation facilities	0.04	1.17	1.62
	Transportation area	1.30	19.058	0.8
	Public facility area	0.32	5.176	0.8
	Agricultural	21.88	0	0
	Forest area	66.71	0	0
	Grassland	0.30	0	0
	Marsh	0.73	0	0
	Bare ground	1.72	0	0
	Water body	2.06	0	0
Child population	0–22	20.59	14.92	0.73
	22–37	20.85	2.57	0.13
	37–60	19.73	5.14	0.27
	60–228	19.68	7.59	0.39
	228–10,747	19.13	69.75	3.65
Elderly population	0–823	20.00	18.01	0.91
	823–1049	20.09	2.31	0.12
	1049–1508	20.08	3.98	0.19
	1508–2252	20.13	5.79	0.29
	2252–31,218	19.68	69.88	3.56
Population density	0–1723	19.04	13.32	0.69
	1723–2872	29.78	4.61	0.16
	2872–4595	22.50	2.90	0.13
	4595–10,913	14.91	8.17	0.55
	10,913–146,455	13.75	70.97	5.17
Distance from hospital (km)	0–6.92	19.32	76.28	3.95
	6.92–12.11	20.99	18.58	0.89
	12.11–17.59	20.15	4.87	0.25
	17.59–24.81	19.90	0.25	0.02
	24.81–73.56	19.62	0	0
Distance from police station (km)	0–3.39	17.89	70.51	3.95
	3.39–5.66	20.88	14.48	0.69
	5.66–7.93	20.16	8.71	0.44
	7.93–11.32	21.22	5.76	0.28
	11.32–72.22	19.83	0.51	0.03
Distance from fire station (km)	0–5.24	18.94	79.35	4.18
	5.24–8.84	20.22	15.64	0.77
	8.84–12.78	20.50	3.84	0.18
	12.78–18.35	20.40	1.15	0.05
	18.35–83.57	19.91	0	0

3.3. Seismic Vulnerability Map

Seismic vulnerability maps were made using the training dataset compiled using the building damage inventory map from the DPM of the Pohang earthquake and applying machine learning algorithms, as discussed above. A combination of 15 seismic-related factors served as the dependent variables, and can mainly be classified as geotechnical, physical, structural, social, and capacity indicators. LogitBoost (Figure 7a), LMT (Figure 7b), and LR (Figure 7b) machine learning algorithms were applied to produce the seismic vulnerability maps. Each pixel in the study area was assigned a specific building damage value using the natural breaks method [7]. The seismic vulnerability maps were classified as safe, low to moderate, high, and very high vulnerability classes.

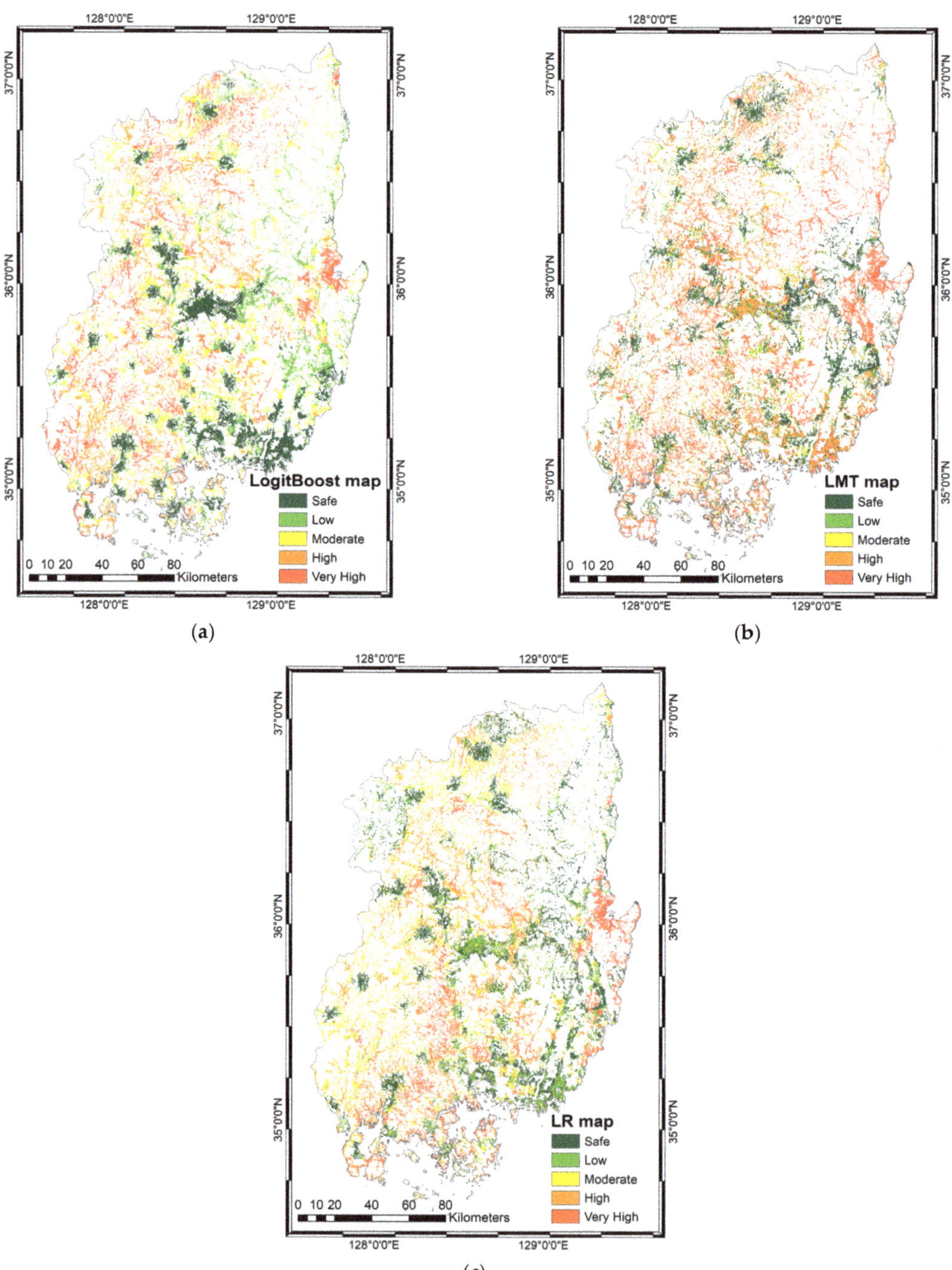

Figure 7. Seismic vulnerability map generated using three algorithms: (**a**) LogitBoost, (**b**) LMT, and (**c**) LR.

Figure 8 shows the distribution of pixels in each seismic vulnerability map generated by LogitBoost, LMT, and LR models. In the LogitBoost model, 29.53% were classified as safe, 17.59% as low risk, 13.18% as moderate risk, 22.95% as high risk, and 16.74% as very high risk. Gyeongju and Pohang were found to be the most vulnerable to earthquake damage. Among big cities in the study area, Busan, Daegu, and Ulsan were classified as safe and low risk. For the LMT model, 19.74% were classified as safe, 18.54% as low risk, 8.25% as moderate risk, 18.27% as high risk, 35.20% as very high risk. Gyeongju and Pohang were found to be the most vulnerable to earthquake damage. Daegu was classified as a safe and low-risk city, while Busan and Ulsan were high and very high risk. In the LR model, 19.91% were classified as safe, 20.08% as low risk, 20.71% as moderate risk, 19.46% as high risk, and 19.85% as very high risk. The distribution of pixels in the low- and high-risk classes in Figure 8 shows similar results for each algorithm, although LMT shows larger numbers of pixels in very high-risk areas. The most vulnerable areas were Gyeongju and Pohang, whereas low and moderate risk areas were Busan and Ulsan. Seismic vulnerability classes were evenly distributed in Daegu.

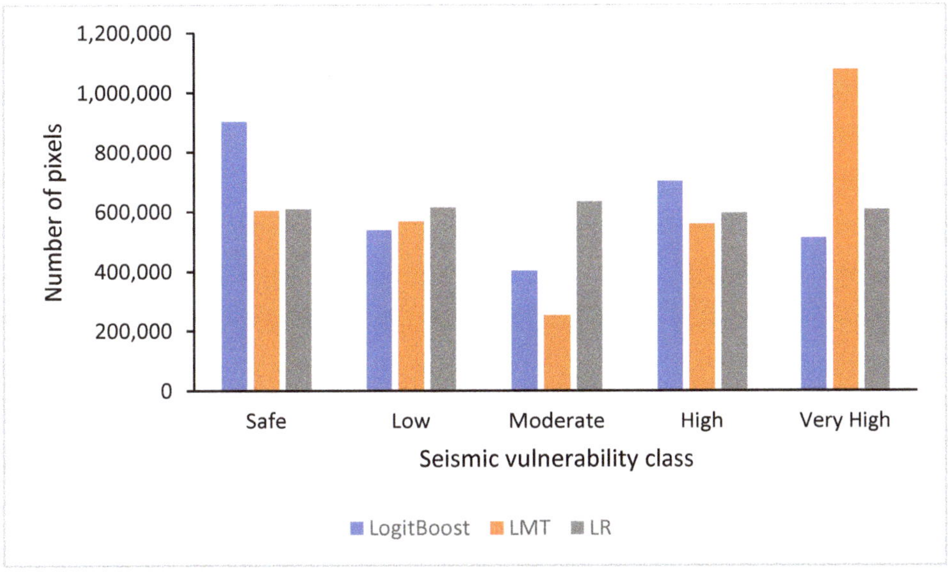

Figure 8. Distribution of pixels in seismic vulnerability classes for LogitBoost, LMT, and LR models.

3.4. Model Validation

A validation step was conducted to assess the reliability of the seismic vulnerability map from each algorithm. ROC curve analysis is a standard way of validating the probability models used to generate seismic vulnerability maps, according to the area under the curve (AUC) [7,14,15]. Higher values indicate more accurate and reliable models. If the AUC, which ranges from 0 to 1, is lower than 0.5, the model is considered unacceptably inaccurate [49]. The accuracy of the seismic vulnerability maps generated using the three algorithms was then evaluated based on ROC curve analysis of the testing dataset (50% of all data). As seen in Figure 9, the AUC values were 0.769, 0.851, and 0.749 for LogitBoost, LMT, and logistic regression, respectively. Thus, the LMT model generated the best seismic vulnerability map in this study. The results indicate that the algorithms are useful to map seismic vulnerability in the southeastern Korean Peninsula. Since all of the AUC values were higher than 0.5, the seismic vulnerability maps produced by all algorithms used in this study are acceptable for predicting vulnerable buildings in southeast Korea [49].

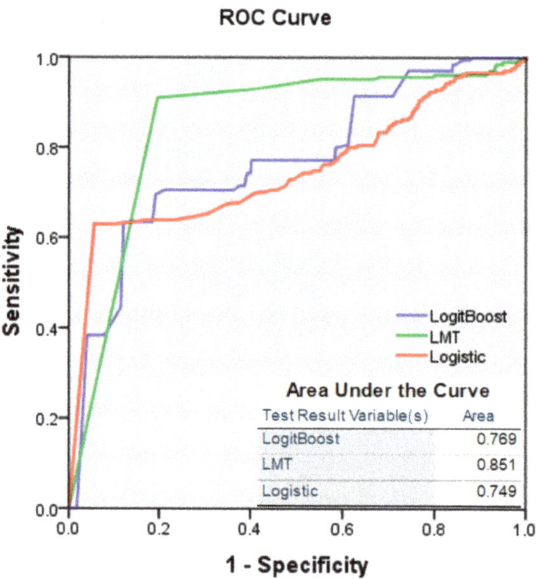

Figure 9. Receiver operating characteristic (ROC) curves associated with seismic vulnerability maps generated by LogitBoost, LMT, and LR algorithms.

4. Discussion

4.1. Building Damage Inventory Map

The building damage inventory map was generated through the DPM method after the Pohang earthquake using Sentinel-1 SAR data. We found a good correlation between the DPM result and the released map from the KMA report about the Pohang earthquake in our qualitative validation. Therefore, the building damage inventory map is reliable. The KMA map was derived from field survey and damage survey data from local governments. The map shows the distribution of the Pohang earthquake's magnitude using the modified Mercalli intensity (MMI) scale [34]. The results were similar to the KMA report that these areas suffered MMI VII to VIII. KMA defines MMI VII to VIII as damage to major structural parts such as pillars, walls, and roofs, even in well-designed and well-built buildings. KMA map also shows some areas that suffered MMI V to VI; however, the DPMs did not show any CODs in these areas. The scales showed that the damage was inside buildings, such as minor cracks in walls and damage caused by dropping objects or tiles; therefore, SAR cannot detect a significant coherence change.

Here, the total damaged areas from DPMs were calculated by multiplying the total number of pixels by the pixel cell size, which for the Gyeongju earthquake yielded an area of 1.09 km^2 and the Pohang earthquake yielded an area of 1.32 km^2. The KMA reported a total of 69 damaged buildings by the Gyeongju earthquake, and 504 damaged buildings by the Pohang earthquake (MMI VII and VIII). The Pohang earthquake caused more damage than the Gyeongju earthquake, some of which was due to the depth of the epicenter. The shallower the epicenter of an earthquake, the more damage it causes. The epicenter depth of the Pohang earthquake was 7 km, while the epicenter depth of the Gyeongju earthquake was 15 km. Additionally, the surface deformation in urban areas with high buildings caused by the Pohang earthquake affected the occurrence of damage in Pohang.

4.2. Seismic Vulnerability Map

Estimating the seismic vulnerability of an area is vital for environmental management and land use planning, among other applications [19]. Although many methods and techniques have been developed to assess earthquake hazards around the world to date, the goals of all these studies are to reduce the economic losses and resulting losses. The method used to create seismic vulnerability maps affects the quality of the mapping. Machine learning techniques are effective. In particular, the method used to generate training and testing data is important. Accurate damage building inventory maps can be obtained using the DPM method; we combined DPM and GIS spatial data to produce an accurate seismic vulnerability map.

The seismic vulnerability maps of all three algorithms used in this study revealed that Pohang and Gyeongju are the most vulnerable areas to earthquake damage. We analyzed seismic-related factors by comparing general patterns of damaged buildings with factor maps. Pohang and Gyeongju are the cities where the two largest recent earthquakes happened. Several factors in these areas had higher values, such as the value of PGA and the distance from the epicenter. Therefore, Gyeongju and Pohang were areas with a high risk of earthquake vulnerability. Buildings within 0–36.90 km of an epicenter corresponded to damaged buildings. This result confirms that most buildings close to epicenters were damaged. The results revealed that areas with a high risk had high population and building density, including Gyeongju and Pohang. Similar to Gyeongju and Pohang, Busan, Daegu, and Ulsan are cities with a high density of buildings and populations. These cities consist of areas of low to moderate risk of seismic vulnerability. One of the main causes of a high risk of seismic damage in countryside areas is wood as a building construction material, while buildings constructed with steel reinforced with concrete have a lower level of seismic vulnerability. Therefore, attention should be paid to reconstruction, retrofitting, or renovation of the buildings in these areas.

Seismic vulnerability maps were validated based on ROC curves and AUC values to assess the accuracy of the maps using testing data, which comprised 50% of the total dataset. The AUC data showed that the LMT algorithm had the highest accuracy of 85.10%, which was 8.2% higher than the LogitBoost algorithm and 10.2% higher than the logistic regression algorithm. Therefore, the LMT algorithm is better to produce seismic vulnerability maps than LogitBoost and logistic regression algorithms.

A survey was conducted with 1256 Korean citizens using a questionnaire of citizen earthquake literacy (CEL) during spring 2020. The survey was based on three dimensions, including citizen knowledge, awareness, and management. We developed 15 questions associated with three dimensions. Table A1 in Appendix A presents the questionnaire that consists of five questions for each dimension, and all questions are configured to be answered on a 5-point Likert scale (strongly agree, agree, neutral, disagree, and strongly disagree). Principal component analysis as an exploratory factor analysis approach was used to examine if the items corresponded to each other and explore the construct validity of the instrument. Cronbach's alpha when an item was deleted was also calculated to see whether the items in the construct reliably measured the same latent variable. Cronbach's alpha values for each dimension were 0.847, 0.822, and 0.849, respectively. Furthermore, Cronbach's alpha when an item was deleted in this study was found to be between 0.77 and 0.84. This result indicated that the survey data used in this study were strongly reliable and consistent [64].

To characterize the profile of participants with higher (or lower) levels of seismic literacy, difference in means analyses using a *t*-test, ANOVA test, and Tukey's honestly significant difference (HSD) post hoc test were carried out. Table 2 shows the average values associated with earthquake literacy, including seismic knowledge, awareness, and management, broken down by sociodemographic characteristics of the participants. The *t*-test result shows that male citizens have higher earthquake literacy but not significantly more than female citizens in all three aspects. The ANOVA test was conducted to determine whether there were differences in the subcategories of seismic literacy depending on age, risk awareness, and final educational background. The results show that participants 20 years of age and below, and 60 years of age and above, declared a higher level of seismic literacy than participants between 30 and 50 years of age. Participants in their 30s declared the lowest level of seismic awareness and participants in their 40s declared the lowest level of seismic knowledge and management among the different age groups. However, there was no significant difference in earthquake literacy in all aspects among all ranges of ages (under 20s to over 60s). Only participants who were aware that where they live is absolutely safe from earthquakes had a statistically significantly higher level of seismic management behavior. They also declared a higher level of seismic knowledge. Participants who were aware that where they live is not safe at all from earthquakes declared a higher level of seismic awareness but declared the lowest level of seismic knowledge and management. This result is an important warning sign for local and regulatory authorities to raise the citizen seismic knowledge and management. In the latter case, participants in graduate school and graduate school graduates declared a higher level of seismic literacy in all three aspects. Participants with an education level of less than high school declared the lowest level of seismic knowledge and high school graduates declared the lowest level of seismic awareness and management.

Table 2. Mean values for seismic literacy.

All Samples (n = 1256)		%	Knowledge		Awareness		Management	
			Mean	SD	Mean	SD	Mean	SD
Gender	Male	50.96	3.1988	0.6655	3.1428	0.6937	3.1759	0.6299
	Female	49.04	2.9032	0.7008	2.9659	0.6591	2.8484	0.6515
Age Group	Under 20	17.75	3.1274	0.7671	3.1229	0.6648	3.1157	0.6535
	30s	19.11	3.0008	0.7141	2.9442	0.7138	2.9450	0.7019
	40s	22.37	2.9993	0.6652	2.9929	0.6592	2.9117	0.6303
	50s	23.49	3.0414	0.6664	3.0569	0.6684	3.0285	0.6372
	Over 60	17.28	3.1244	0.6862	3.1917	0.6886	3.1060	0.6692
Risk Awareness	No, not at all	6.13	2.9974	0.7877	3.1922	0.7756	2.9299	0.7812
	No, hardly not	29.30	3.0538	0.7050	3.0946	0.7120	2.9533	0.6851
	It's normal	41.64	3.0164	0.6718	3.0191	0.6340	2.9897	0.6121
	Yes, to some extent	20.46	3.1354	0.6694	3.0412	0.6471	3.1447	0.6095
	Yes, absolutely	2.47	3.1484	0.9946	3.0065	1.0462	3.3226	0.9888
Final Education	Under high school	1.11	2.7857	0.8282	3.1000	0.8727	3.0571	0.8821
	High school graduate	16.16	2.8512	0.6770	2.9586	0.6560	2.8975	0.6496
	In college (or university)	5.81	3.3151	0.6960	3.3014	0.6482	3.2877	0.5588
	College (or university) graduate	5.81	3.0227	0.6805	3.0336	0.6759	2.9755	0.6516
	In graduate school	64.41	3.4762	0.6999	3.1048	0.6888	3.3524	0.5759
	Graduate school graduate	10.83	3.3632	0.6685	3.1912	0.7176	3.2250	0.6818

Further analysis was conducted to find significant differences through Tukey's HSD post hoc test and showed that citizens who were aware that where they live is absolutely safe from earthquakes had a statistically significantly higher level of seismic management behavior. Participants with a highly educated background had the highest level of seismic knowledge.

The participants of 60 years old and above declared the highest level of seismic literacy. Some authors posit that this could be explained because adults in this stage of life have acquired greater experience and care responsibilities (either for others or their assets), which may give rise to increased interest in involving themselves in preparedness measures [65]. Participants' educational background also influenced their seismic literacy, especially in seismic knowledge, in line with previous studies [66,67]. Groups with less seismic literacy should be a target of intervention in order to raise earthquake risk awareness and motivate them to adopt preparedness actions.

Nevertheless, in general, the participants declared relatively low earthquake literacy, including seismic knowledge, awareness, and preparedness. This means that they are not properly prepared for earthquake hazards. For the areas that were classified as highly vulnerable to earthquakes, the focus should be on building reconstruction and retrofitting [16], whereas for those in moderate-risk areas, the focus should be on building renovation to reduce their seismic vulnerability. Areas with high building and population densities, such as Daegu, Busan, and Ulsan, should develop education programs to improve earthquake risk awareness and seismic literacy. The government could increase earthquake literacy, starting with a conscious approach to Korean citizens, especially in high vulnerability areas.

Finally, the institution responsible for developing local disaster risk reduction plans and programs should appropriately characterize their target audiences and areas if they expect to obtain more effective and efficient results. We expect that the results reported in this study will be useful input to achieve this. However, this study has certain limitations. First, the questionnaire could provide more information of the participants profile such as marital status, children in the house, annual household income, and household type and could be analyzed to provide specific information about participants' preparedness. Future studies could investigate the interaction between these variables to find specific patterns of seismic literacy.

5. Conclusions

We conducted a survey of 1256 participants to investigate the seismic literacy of Korean citizens, including seismic knowledge, awareness, and management, using a questionnaire of citizen earthquake literacy (CEL) following the 2017 Pohang earthquake. The results declared that the citizens had low literacy, which means that they are not properly prepared for earthquake hazards. To develop an earthquake risk reduction plan and program efficiently and effectively, not only must one appropriately characterize the target audience, but also indicate high potential earthquake zones and potential earthquake damage. Therefore, this study mapped and analyzed the seismic vulnerability in southeast Korea using LogitBoost, logistic model tree (LMT), and logistic regression (LR) machine learning algorithms based on a building damage inventory map. The building damage locations were generated after the 2017 Pohang earthquake using the damage proxy map (DPM) method from the Sentinel-1 synthetic aperture radar (SAR) data. The DPMs manage to detect coherence loss, which indicates damaged buildings in residential and commercial areas due to the Pohang earthquake and show a good correlation with the Korea Meteorological Administration (KMA) report with modified Mercalli intensity (MMI) scale values of more than VII (seven). The damage locations were randomly divided into two datasets: 50% for training the vulnerability models and 50% for validating the models in terms of accuracy and reliability. Fifteen seismic-related factors were used to construct a model for each algorithm. Model validation based on the area under the receiver operating curve (AUC) was used to determine model accuracy. The AUC values of seismic vulnerability maps using the LogitBoost, LMT, and LR algorithms were 0.769, 0.851, and 0.749, respectively. We suggest that earthquake preparedness efforts should focus on reconstruction, retrofitting, renovation, and seismic education in areas with high seismic vulnerability in South Korea. The results of this study are expected to be beneficial for engineers and policymakers aiming at developing disaster risk reduction plans, policies, and programs due to future seismic activity in South Korea.

Author Contributions: Conceptualization, C.-W.L. and K.-Y.L.; methodology, M.S. and M.H.; software, M.S. and M.H.; validation. K.-Y.L., M.S. and J.H.; formal analysis, C.-W.L.; investigation, J.H. and M.H.; resources, M.S.; data curation, M.S., M.H. and J.H.; writing—original draft preparation, A.S.N.; writing—review and editing, M.S., A.S.N. and J.H.; visualization, M.S.; supervision, K.-Y.L.; project administration, J.H.; funding acquisition, K.-Y.L. All authors have read and agreed to the published version of the manuscript.

Funding: This research was supported by a grant from the National Research Foundation of Korea provided by the government of Korea (No. 2019R1A2C1085686) and also supported by a 2020 Research Grant from Kangwon National University.

Institutional Review Board Statement: Not applicable.

Informed Consent Statement: Informed consent was obtained from all subjects involved in the study.

Data Availability Statement: The dataset of seismic-related factor can be found on: https://www.bigdata-environment.kr/user/main.do (accessed on 31 March 2021). Sentinel-1 A/B SAR images are freely available by European Space Agency and distributed and archived by Alaska Satellite Facility (https://search.asf.alaska.edu/#/ (accessed on 31 March 2021)). The DEM SRTM data and epicenter location used in this study were obtained from the United States Geological Survey (USGS) (https://earthexplorer.usgs.gov/ (accessed on 31 March 2021) and https://earthquake.usgs.gov/earthquakes/map (accessed on 31 March 2021)).

Conflicts of Interest: The authors declare no conflict of interest.

Appendix A

Table A1. Survey questions asked by a researcher to prompt discussion.

No	Knowledge	1	2	3	4	5
1.	I know why earthquakes occur	1.91	11.94	42.68	38.14	5.33
2.	I can distinguish between earthquake magnitude and scale	4.70	25.16	37.02	29.22	3.90
3.	I can explain the terms related to earthquakes (e.g., epicenter, hypocenter, main earthquake, aftershock)	5.89	25.51	38.06	25.08	4.46
4.	I do not have problem understanding earthquake news or articles	0.64	10.27	39.01	42.36	7.72
5.	I know how and what technology is used to study earthquakes	11.39	41.24	36.15	9.00	2.23
	Awareness					
6.	I know the shelter close to my home for escaping from earthquakes	5.49	29.94	35.67	25.32	3.58
7.	I know prevention items when earthquakes occur	8.44	42.68	32.96	14.01	1.91
8.	I know the earthquake will affect to my area and social community	2.15	14.09	33.52	42.12	8.12
9.	I know what to do when earthquakes happen	0.88	11.86	38.14	43.15	5.97
10.	I know the earthquake early warning service	4.38	26.27	40.21	26.51	2.63
	Management					
11.	I have a calm attitude for earthquakes	2.55	21.66	49.28	23.73	2.79
12.	I can handle terror, fear in earthquake situations	3.66	26.04	46.10	21.74	2.47
13.	I changed my house to reduce damage, e.g., falling furniture or broken glass, when earthquakes happen	8.04	37.98	35.67	16.48	1.83
14.	I can rapidly follow the earthquake early warning (message) service	2.23	13.93	45.70	34.08	4.06
15.	I can return to daily life after an earthquake	1.83	13.38	49.12	32.40	3.26

1. Yes, absolutely; 2. Yes; 3. It's normal; 4. No; 5. Absolutely not.

References

1. Kim, Y.S.; Park, J.Y.; Kim, J.H.; Shin, H.C.; Sanderson, D.J. Thrust geometries in unconsolidated Quaternary sediments and evolution of the Eupchon Fault, southeast Korea. *Isl. Arc* **2004**, *13*, 403–415. [CrossRef]
2. Lee, K.; Yang, W.S. Historical seismicity of Korea. *Bull. Seismol. Soc. Am.* **2006**, *96*, 846–855. [CrossRef]
3. Kang, S.; Kim, B.; Bae, S.; Lee, H.; Kim, M. Earthquake-Induced Ground Deformations in the Low-Seismicity Region: A Case of the 2017 M5.4 Pohang, South Korea, Earthquake. *Earthq. Spectra* **2019**, *35*, 1235–1260. [CrossRef]
4. Grigoli, F.; Cesca, S.; Rinaldi, A.P.; Manconi, A.; López-Comino, J.A.; Clinton, J.F.; Westaway, R.; Cauzzi, C.; Dahm, T.; Wiemer, S. The November 2017 M_w 5.5 Pohang earthquake: A possible case of induced seismicity in South Korea. *Science* **2018**, *360*, 1003–1006. [CrossRef] [PubMed]
5. Woo, J.-U.; Kim, M.; Sheen, D.-H.; Kang, T.-S.; Rhie, J.; Grigoli, F.; Ellsworth, W.L.; Giardini, D. An In-Depth Seismological Analysis Revealing a Causal Link between the 2017 M_W 5.5 Pohang Earthquake and EGS Project. *J. Geophys. Res. Solid Earth* **2019**, *124*, 13060–13078. [CrossRef]
6. Lee, K.-K. *Summary Report of the Korean Government Commision on Relation between the 2017 Pohang Earthquake and EGS Project*; The Geological Society of Korea: Seoul, Korea, 2019.
7. Han, J.; Kim, J.; Park, S.; Son, S.; Ryu, M. Seismic Vulnerability Assessment and Mapping of Gyeongju, South Korea Using Frequency Ratio, Decision Tree, and Random Forest. *Sustainability* **2020**, *12*, 7787. [CrossRef]
8. Sang-Sun toward Better Risk Preparedness for Cultural Heritage. Available online: http://www.koreanheritage.kr//inside/view.jsp?articleNo=15 (accessed on 19 January 2021).
9. Doo, R. Artifacts Damaged in Record-Breaking Earthquake in Gyeongju. Available online: https://www.thejakartapost.com/life/2016/09/19/artifacts-damaged-in-record-breaking-earthquake-in-gyeongju.html (accessed on 18 January 2021).
10. Werner, C.; Wegmüller, U.; Strozzi, T.; Wiesmann, A. GAMMA SAR and interferometric processing software. In Proceedings of the ERS—ENVISAT Symposium, Gothenburg, Sweden, 16–20 October 2000; Special Publication ESA SP. European Space Agency: Paris, France, 2000; pp. 211–219.
11. Zebker, H.A.; Goldstein, R.M. Topographic Mapping from Interferometric Synthetic Aperture Radar Observations. In *Digest—International Geoscience and Remote Sensing Symposium (IGARSS)*; IEEE: New York, NY, USA, 1985; pp. 113–117.
12. Panahi, M.; Rezaie, F.; Meshkani, S.A. Seismic vulnerability assessment of school buildings in Tehran city based on AHP and GIS. *Nat. Hazards Earth Syst. Sci. Discuss.* **2013**, *1*, 4511–4538. [CrossRef]
13. Jena, R.; Pradhan, B.; Beydoun, G. Earthquake vulnerability assessment in Northern Sumatra province by using a multi-criteria decision-making model. *Int. J. Disaster Risk Reduct.* **2020**, *46*, 101518. [CrossRef]
14. Yariyan, P.; Avand, M.; Soltani, F.; Ghorbanzadeh, O.; Blaschke, T. Earthquake vulnerability mapping using different hybrid models. *Symmetry* **2020**, *12*, 405. [CrossRef]
15. Han, J.; Park, S.; Kim, S.; Son, S.; Lee, S.; Kim, J. Performance of Logistic Regression and Support Vector Machines for Seismic Vulnerability Assessment and Mapping: A Case Study of the 12 September 2016 ML5.8 Gyeongju Earthquake, South Korea. *Sustainability* **2019**, *11*, 7038. [CrossRef]
16. Alizadeh, M.; Hashim, M.; Alizadeh, E.; Shahabi, H.; Karami, M.; Beiranvand Pour, A.; Pradhan, B.; Zabihi, H. Multi-Criteria Decision Making (MCDM) Model for Seismic Vulnerability Assessment (SVA) of Urban Residential Buildings. *ISPRS Int. J. Geo-Inf.* **2018**, *7*, 444. [CrossRef]
17. Han, J.; Kim, J. GIS-Based Seismic Vulnerability Mapping and Assessment Using AHP: A Case Study of Gyeongju, Korea. *Korean J. Remote Sens.* **2019**, *35*, 217–228.
18. Rezaie, F.; Panahi, M. GIS modeling of seismic vulnerability of residential fabrics considering geotechnical, structural, social and physical distance indicators in Tehran using multi-criteria decision-making techniques. *Nat. Hazards Earth Syst. Sci.* **2015**, *15*, 461–474. [CrossRef]
19. Lee, S.; Panahi, M.; Pourghasemi, H.R.; Shahabi, H.; Alizadeh, M.; Shirzadi, A.; Khosravi, K.; Melesse, A.M.; Yekrangnia, M.; Rezaie, F.; et al. SEVUCAS: A Novel GIS-Based Machine Learning Software for Seismic Vulnerability Assessment. *Appl. Sci.* **2019**, *9*, 3495. [CrossRef]
20. Yun, S.-H.; Hudnut, K.; Owen, S.; Webb, F.; Simons, M.; Sacco, P.; Gurrola, E.; Manipon, G.; Liang, C.; Fielding, E.; et al. Rapid Damage Mapping for the 2015 M_w 7.8 Gorkha Earthquake Using Synthetic Aperture Radar Data from COSMO–SkyMed and ALOS-2 Satellites. *Seismol. Res. Lett.* **2015**, *86*, 1549–1556. [CrossRef]
21. Tay, C.W.J.; Yun, S.-H.; Chin, S.T.; Bhardwaj, A.; Jung, J.; Hill, E.M. Rapid flood and damage mapping using synthetic aperture radar in response to Typhoon Hagibis, Japan. *Sci. Data* **2020**, *7*, 100–108. [CrossRef] [PubMed]
22. Hough, S.E.; Yun, S.-H.; Jung, J.; Thompson, E.; Parker, G.A.; Stephenson, O. Near-Field Ground Motions and Shaking from the 2019 Mw 7.1 Ridgecrest, California, Mainshock: Insights from Instrumental, Macroseismic Intensity, and Remote-Sensing Data. *Bull. Seismol. Soc. Am.* **2020**, *110*, 1506–1516. [CrossRef]
23. Biass, S.; Jenkins, S.; Lallemant, D.; Lim, T.N.; Williams, G.; Yun, S.-H. Remote sensing of volcanic impacts. In *Forecasting and Planning for Volcanic Hazards, Risks, and Disasters*; Elsevier: Amsterdam, The Netherlands, 2021; pp. 473–491.
24. Choe, B.-N. *The National Atlas of Korea*; National Geography Information Institute: Suwon, Korea, 2019.
25. Korean Statistical Information Service Statistics Korea. Population Census. Available online: https://kosis.kr/statHtml/statHtml.do?orgId=101&tblId=DT_1IN1502&conn_path=I2&language=en (accessed on 21 January 2020).

26. Kim, H.-S.; Sun, C.-G.; Cho, H.-I. Geospatial Assessment of the Post-Earthquake Hazard of the 2017 Pohang Earthquake Considering Seismic Site Effects. *ISPRS Int. J. Geo-Inf.* **2018**, *7*, 375. [CrossRef]

27. Kim, Y.; Rhie, J.; Kang, T.-S.; Kim, K.-H.; Kim, M.; Lee, S.-J. The 12 September 2016 Gyeongju earthquakes: 1. Observation and remaining questions. *Geosci. J.* **2016**, *20*, 747–752. [CrossRef]

28. Jin, K.; Lee, J.; Lee, K.-S.; Kyung, J.B.; Kim, Y.-S. Earthquake damage and related factors associated with the 2016 ML = 5.8 Gyeongju earthquake, southeast Korea. *Geosci. J.* **2020**, *24*, 141–157. [CrossRef]

29. Korea Meteorological Administration. *2018 Earthquake Annual Report*; Korea Meteorological Administration: Seoul, Korea, 2019.

30. Korea Meteorological Administration. *2019 Earthquake Annual Report*; Korea Meteorological Administration: Seoul, Korea, 2020.

31. MML, S.; Mote, T.; Pappin, J. Seismic hazard assessment of South Korea. In Proceedings of the Japanese Geotechnical Society Special Publication; Japanese Geotechnical Society: Tokyo, Japan, 2000; pp. 755–760.

32. Ellsworth, W.L.; Giardini, D.; Townend, J.; Ge, S.; Shimamoto, T. Triggering of the Pohang, Korea, Earthquake (Mw 5.5) by enhanced geothermal system stimulation. *Seismol. Res. Lett.* **2019**, *90*, 1844–1858. [CrossRef]

33. Zebker, H.A.; Villasenor, J. Decorrelation in interferometric radar echoes. *IEEE Trans. Geosci. Remote Sens.* **1992**, *30*, 950–959. [CrossRef]

34. Korea Meteorological Administration. *Pohang Eartquake Analysis Report*; Korea Meteorological Administration: Seoul, Korea, 2018.

35. Yariyan, P.; Karami, M.R.; Ali Abbaspour, R. Exploitation of Mcda to Learn the Radial Base Neural Network (rbfnn) Aim Physical and Social Vulnerability Analysis Versus the Earthquake (Case Study: Sanandaj City, Iran). *ISPRS Int. Arch. Photogramm. Remote Sens. Spat. Inf. Sci.* **2019**, *XLII-4/W18*, 1071–1078. [CrossRef]

36. Pachauri, A.K.; Pant, M. Landslide hazard mapping based on geological attributes. *Eng. Geol.* **1992**, *32*, 81–100. [CrossRef]

37. Kim, K.-H.; Kang, T.-S.; Rhie, J.; Kim, Y.; Park, Y.; Kang, S.Y.; Han, M.; Kim, J.; Park, J.; Kim, M.; et al. The 12 September 2016 Gyeongju earthquakes: 2. Temporary seismic network for monitoring aftershocks. *Geosci. J.* **2016**, *20*, 753–757. [CrossRef]

38. Kang, T.-S.; Kim, D.K. *Convergence Research Review*; Convergence Research Policy Center: Seoul, Korea, 2017.

39. Sivakumar, N.; Karthik, S.; Thangaraj, S.; Saravanan, S.; Shidhardhan, C.K. Seismic Vulnerability of Open Ground Floor Columns in Multi Storey Buildings. *Int. J. Sci. Eng. Res.* **2013**, *1*, 52–58.

40. Cole, G.L.; Dhakal, R.P.; Turner, F.M. Building pounding damage observed in the 2011 Christchurch earthquake. *Earthq. Eng. Struct. Dyn.* **2012**, *41*, 893–913. [CrossRef]

41. Rimal, B.; Baral, H.; Stork, N.; Paudyal, K.; Rijal, S. Growing City and Rapid Land Use Transition: Assessing Multiple Hazards and Risks in the Pokhara Valley, Nepal. *Land* **2015**, *4*, 957–978. [CrossRef]

42. Hassanzadeh, R.; Nedović-Budić, Z.; Alavi Razavi, A.; Norouzzadeh, M.; Hodhodkian, H. Interactive approach for GIS-based earthquake scenario development and resource estimation (Karmania hazard model). *Comput. Geosci.* **2013**, *51*, 324–338. [CrossRef]

43. Rahman, N.; Ansary, M.A.; Islam, I. GIS based mapping of vulnerability to earthquake and fire hazard in Dhaka city, Bangladesh. *Int. J. Disaster Risk Reduct.* **2015**, *13*, 291–300. [CrossRef]

44. Armaş, I.; Toma-Danila, D.; Ionescu, R.; Gavriş, A. Vulnerability to Earthquake Hazard: Bucharest Case Study, Romania. *Int. J. Disaster Risk Sci.* **2017**, *8*, 182–195. [CrossRef]

45. Karimzadeh, S.; Miyajima, M.; Hassanzadeh, R.; Amiraslanzadeh, R.; Kamel, B. A GIS-based seismic hazard, building vulnerability and human loss assessment for the earthquake scenario in Tabriz. *Soil Dyn. Earthq. Eng.* **2014**, *66*, 263–280. [CrossRef]

46. Pradhan, B.; Abokharima, M.H.; Jebur, M.N.; Tehrany, M.S. Land subsidence susceptibility mapping at Kinta Valley (Malaysia) using the evidential belief function model in GIS. *Nat. Hazards* **2014**, *73*, 1019–1042. [CrossRef]

47. Hakim, W.; Achmad, A.; Lee, C.-W. Land Subsidence Susceptibility Mapping in Jakarta Using Functional and Meta-Ensemble Machine Learning Algorithm Based on Time-Series InSAR Data. *Remote Sens.* **2020**, *12*, 3627. [CrossRef]

48. Moung-Jin, L.; Won-Kyong, S.; Joong-Sun, W.; Inhye, P.; Saro, L. Spatial and temporal change in landslide hazard by future climate change scenarios using probabilistic-based frequency ratio model. *Geocarto Int.* **2014**, *29*, 639–662. [CrossRef]

49. Fawcett, T. An introduction to ROC analysis. *Pattern Recognit. Lett.* **2006**, *27*, 861–874. [CrossRef]

50. Friedman, J.; Hastie, T.; Tibshirani, R. Additive logistic regression: A statistical view of boosting (with discussion and a rejoinder by the authors). *Ann. Stat.* **2000**, *28*, 337–407. [CrossRef]

51. Quinlan, J.R. *C4. 5: Programs for Machine Learning*; Elsevier: Amsterdam, The Netherlands, 2014; ISBN 0080500587.

52. Landwehr, N.; Hall, M.; Frank, E. Logistic Model Trees. *Mach. Learn.* **2005**, *59*, 161–205. [CrossRef]

53. Lim, T.-S.; Loh, W.-Y.; Shih, Y.-S. A Comparison of Prediction Accuracy, Complexity, and Training Time of Thirty-Three Old and New Classification Algorithms. *Mach. Learn.* **2000**, *40*, 203–228. [CrossRef]

54. Shahabi, H.; Ahmad, B.B.; Khezri, S. Evaluation and comparison of bivariate and multivariate statistical methods for landslide susceptibility mapping (case study: Zab basin). *Arab. J. Geosci.* **2013**, *6*, 3885–3907. [CrossRef]

55. Doetsch, P.; Buck, C.; Golik, P.; Hoppe, N.; Kramp, M.; Laudenberg, J.; Oberdörfer, C.; Steingrube, P.; Forster, J.; Mauser, A. Logistic model trees with auc split criterion for the kdd cup 2009 small challenge. In Proceedings of the KDD-Cup 2009 Competition. *PMLR Proc. Mach. Learn. Res.* **2009**, *7*, 77–88.

56. McFadden, D. Conditional logit analysis of qualitative choice behavior. In *Frontiers in Econometrics*; Academic Press: Cambridge, MA, USA, 1973.

57. Kleinbaum, D.G.; Klein, M. *Survival Analysis*; Springer: New York, NY, USA, 2010; ISBN 1441966455.

58. Colkesen, I.; Sahin, E.K.; Kavzoglu, T. Susceptibility mapping of shallow landslides using kernel-based Gaussian process, support vector machines and logistic regression. *J. Afr. Earth Sci.* **2016**, *118*, 53–64. [CrossRef]
59. Lee, S. Application of logistic regression model and its validation for landslide susceptibility mapping using GIS and remote sensing data. *Int. J. Remote Sens.* **2005**, *26*, 1477–1491. [CrossRef]
60. Wang, L.-J.; Guo, M.; Sawada, K.; Lin, J.; Zhang, J. A comparative study of landslide susceptibility maps using logistic regression, frequency ratio, decision tree, weights of evidence and artificial neural network. *Geosci. J.* **2016**, *20*, 117–136. [CrossRef]
61. Bui, D.T.; Pradhan, B.; Lofman, O.; Revhaug, I.; Dick, O.B. Landslide susceptibility assessment in the Hoa Binh province of Vietnam: A comparison of the Levenberg–Marquardt and Bayesian regularized neural networks. *Geomorphology* **2012**, *171*, 12–29.
62. Xu, C.; Xu, X.; Dai, F.; Saraf, A.K. Comparison of different models for susceptibility mapping of earthquake triggered landslides related with the 2008 Wenchuan earthquake in China. *Comput. Geosci.* **2012**, *46*, 317–329. [CrossRef]
63. Han, M. The 6th Day of the Earthquake . . . Gyeongju Citizens "Continuously Anxious over Aftershocks". Available online: https://www.yna.co.kr/view/AKR20160917017800053 (accessed on 21 January 2021).
64. Gliem, J.A.; Gliem, R.R. Calculating, Interpreting, and Reporting Cronbach's Alpha Reliability Coefficient for Likert-Type Scales. In *Midwest Rest to Practice Conference in Adult Continuing and Community Education*; The Ohio State University: Colombus, OH, USA, 2003; p. 87.
65. Becker, J.S.; Paton, D.; Johnston, D.M.; Ronan, K.R.; McClure, J. The role of prior experience in informing and motivating earthquake preparedness. *Int. J. Disaster Risk Reduct.* **2017**, *22*, 179–193. [CrossRef]
66. Paul, B.K.; Bhuiyan, R.H. Urban earthquake hazard: Perceived seismic risk and preparedness in Dhaka City, Bangladesh. *Disasters* **2010**, *34*, 337–359. [CrossRef] [PubMed]
67. Tekeli-Yeşil, S.; Dedeoğlu, N.; Tanner, M.; Braun-Fahrlaender, C.; Obrist, B. Individual preparedness and mitigation actions for a predicted earthquake in Istanbul. *Disasters* **2010**, *34*, 910–930. [CrossRef]

 *remote sensing*

Article

Land Subsidence Susceptibility Mapping in Jakarta Using Functional and Meta-Ensemble Machine Learning Algorithm Based on Time-Series InSAR Data

Wahyu Luqmanul Hakim, **Arief Rizqiyanto Achmad** and **Chang-Wook Lee** *

Division of Science Education, Kangwon National University, Gangwon-do, Chuncheon-si 24341, Korea;
wahyulhakim@kangwon.ac.kr (W.L.H.); ariefrizqiyanto@kangwon.ac.kr (A.R.A.)
* Correspondence: cwlee@kangwon.ac.kr

Received: 10 September 2020; Accepted: 3 November 2020; Published: 4 November 2020

Abstract: Areas at risk of land subsidence in Jakarta can be identified using a land subsidence susceptibility map. This study evaluates the quality of a susceptibility map made using functional (logistic regression and multilayer perceptron) and meta-ensemble (AdaBoost and LogitBoost) machine learning algorithms based on a land subsidence inventory map generated using the Sentinel-1 synthetic aperture radar (SAR) dataset from 2017 to 2020. The land subsidence locations were assessed using the time-series interferometry synthetic aperture radar (InSAR) method based on the Stanford Method for Persistent Scatterers (StaMPS) algorithm. The mean vertical deformation maps from ascending and descending tracks were compared and showed a good correlation between displacement patterns. Persistent scatterer points with mean vertical deformation value were randomly divided into two datasets: 50% for training the susceptibility model and 50% for validating the model in terms of accuracy and reliability. Additionally, 14 land subsidence conditioning factors correlated with subsidence occurrence were used to generate land subsidence susceptibility maps from the four algorithms. The receiver operating characteristic (ROC) curve analysis showed that the AdaBoost algorithm has higher subsidence susceptibility prediction accuracy (81.1%) than the multilayer perceptron (80%), logistic regression (79.4%), and LogitBoost (79.1%) algorithms. The land subsidence susceptibility map can be used to mitigate disasters caused by land subsidence in Jakarta, and our method can be applied to other study areas.

Keywords: Jakarta; land subsidence susceptibility mapping; time-series InSAR; StaMPS processing; machine learning

1. Introduction

Several cities in Indonesia suffer from degradation at the ground level of buildings, known as land subsidence [1,2]. In Jakarta, this process has had a severe impact on urban infrastructure, leading to cracks in buildings, roads and damage to drainage systems [3]. These conditions are problematic because land subsidence may expand coastal flood areas due to sea-level rise [4]. Heavy monsoon rainfall [5] has caused frequent river flooding; if this occurs again, Jakarta could be submerged entirely underwater [1,4].

Recent studies of land subsidence in Jakarta have used various geodetic measurement methods, such as leveling surveys [3] and a global positioning system (GPS) surveys [6,7]. These studies indicated that excessive groundwater extraction is the leading cause of land subsidence and compaction to the vulnerable aquifer system [8]. This compaction may be exacerbated by natural consolidation since Jakarta's landform mostly comprises young alluvial soils that cannot support the weight of human-made structures [9,10]. Therefore, it is essential to monitor land subsidence in Jakarta to predict further possible occurrences and mitigate damage [11,12].

Over the last decade, land subsidence susceptibility maps have been generated using geological, geomorphological, topographical, and hydrological data; these are considered the main factors influencing land subsidence [11,13,14]. Various methods are used to generate land subsidence susceptibility maps, including frequency ratios (FR) [12,15], weight of evidence (WOE) [16,17], logistic regression (LR) [18], evidential belief functions [11], analytical hierarchy processes (AHP) [19], support vector machines (SVM) [14,20], decision trees [21,22], fuzzy logic [12,23], adaptive neuro-fuzzy inference systems (ANFIS) [24,25], and artificial neural networks (ANN) [26,27]. In general, using a single modeling method leads to lower predictive accuracy than an ensemble method that uses a combination of models and machine learning algorithms [13,28]. Machine learning algorithms have the advantage of finding unpredictable relationships in datasets at multiple scales and have, thus, been recommended to obtain accurate land subsidence susceptibility maps [22].

Another challenge in generating land subsidence susceptibility maps is the low availability of land subsidence inventory maps. In this study, a land subsidence inventory map was generated via time-series interferometry synthetic aperture radar (InSAR) analysis. This technique can be applied to measure the displacement of the earth's surface with an accuracy of up to millimeters per year by improving the selection of coherent pixels and reducing atmospheric noise [29–33]. It has been widely used to monitor land subsidence and generate land subsidence maps, for example, in large cities in Mexico [33,34], Kurdistan, and Iran [35], in Yuncheng Basin, China [36], and in coastal cities and areas such as Venice, Italy [37], New Orleans, United States [38], and Shanghai, China [39]. The InSAR algorithm has been successfully applied to earthquakes [40], volcanic activities [41,42], crustal deformation [31], landslides [43], manmade deformations [44], excessive groundwater extraction [3], mining activities [45], and natural consolidation of young alluvial soil [9].

The recent studies of monitoring land subsidence in Jakarta using interferometry synthetic aperture radar (InSAR) techniques was conducted using the Small Baseline Subset (SBAS) algorithm from 2007 to 2009 [1] and using Geodesy and Earth Observing Systems-Persistent Scatterer Interferometry (GEOS-PSI) algorithm from 2007 to 2010 [46]. Both studies utilized Advanced Land Observing Satellite (ALOS) phased array type-L synthetic aperture radar (PALSAR) data to produce a land subsidence map, and both compared their results with GPS survey data. However, Jakarta's land subsidence map requires updating as it has remained unchanged for the past 10 years, and research on land subsidence susceptibility maps was not found in Jakarta.

Therefore, this study's objective was to generate the updated land subsidence map in Jakarta using the Stanford Method for Persistent Scatterers (StaMPS) with the Sentinel-1 synthetic aperture radar (SAR) dataset from March 2017 to May 2020 in ascending track and March 2017 to April 2020 in a descending track. The mean vertical deformation maps in ascending and descending tracks were compared to validate Jakarta's land subsidence location. After that, the land subsidence map obtained with this method was used as an inventory map to generate a land subsidence susceptibility map in Jakarta that predicted the areas at risk of land subsidence in the future. Two meta-ensemble machine learning algorithms (adaptive boosting (AdaBoost) and Logit Boost), and two functional machine learning algorithms (logistic regression and multilayer perceptron) were used. The result of the land subsidence susceptibility map produced by these algorithms was evaluated to compare all models' accuracy and reliability using receiver operating characteristic (ROC) curve analysis.

2. Study Area

Jakarta is Indonesia's capital city (Figure 1a), located at 6°17′ south (S) and 106°82′ east (E), on the northern coast of western Java. It is considered a lowland area, with an average altitude of ±7 m above sea level [3,47]. In 2019, its population was 10.5 million, with a population growth rate of about 1.19% per year. The population density was 15,900 people per km^2, within a total land area of 662.33 km^2 [47]. Historically, the population of Jakarta gradually migrated to the other districts and municipalities that are part of the Jakarta Metropolitan Region (JMR), which covers a total area of 5897 km^2 and includes Bogor, Depok, Tangerang, and Bekasi [48] (Figure 1b). This migration to Jakarta's outskirts led to increased urban development that could cause land subsidence in these areas [3]. Our study area covers the JMR; however, for simplicity, we refer to it as Jakarta. The three administrative areas were chosen as study area due to land subsidence reported on those areas from the recent study using GPS surveys, leveling surveys, and InSAR techniques. Training and testing data in Figure 1b used in this study to generate a land subsidence susceptibility map were assessed from the land subsidence inventory map.

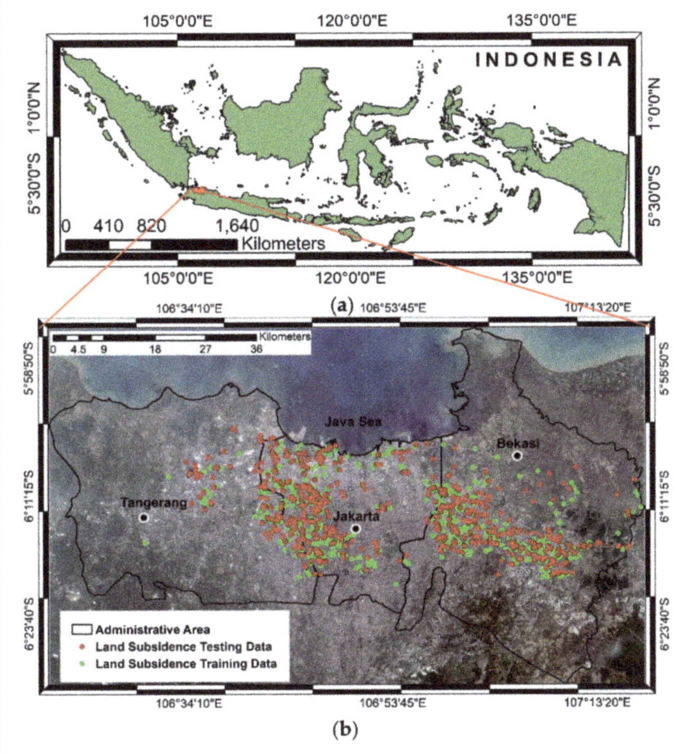

Figure 1. (**a**) Study area location in Indonesia and (**b**) training and testing data sets from land subsidence location within Jakarta Metropolitan Region (JMR) depicted from Sentinel-2 satellite imagery taken on 28 August 2020, divided into three administrative areas: Jakarta, Tangerang, and Bekasi.

The geological and geomorphological area in Jakarta lithologically was dominated by alluvium landform (50.20%) and alluvium fans (19.66%), followed by Serpong form (7.03%) that dominated by fragmented pumice sandstones, some limestone, and andesite. A volcanic pyroclastic flow formed tuff Banten (6.77%) and upper Banten tuff (4.88%), Sandstone unit (4.05%), swamp deposit (1.75%), Cihoe form (1.61%), beach ridge deposit (1.53%), and old alluvium (0.74%). Subang form (0.64%) was dominated by layered claystone lithology with limestone and marl found locally, marine deposit (0.52%), and young volcanic rocks (0.22%); Bojongmanik form (0.22%) was dominated by alternating sandstones and claystone inserted by limestones and coastal deposit (0.13%); Parigi form (0.05%) was dominated by medium limestone, lake (0.01%), and sandstone tuff (0.01%). The domination of alluvium landform has a risk of land subsidence due to the compaction of natural consolidation worsened by a human-made structure [3,9] covering the Jakarta area's land use. Land use in Jakarta was dominated by settlement area with 45.72% and followed by 39.79% of rice field, 5.94% of dryland agriculture, 4.75% of fish pond, 1.98% of shrub-mixed dryland farms, 0.84% of airport, 0.51% of swamp, 0.35% of estate crop plantation, 0.09% of barren land, 0.03% of secondary mangrove forest, and 0.004% of swamp shrub.

Research on land subsidence in Jakarta has been conducted since 1982 using leveling surveys. There were two main monitoring periods: (1) 1982–1991 and (2) 1991–1997. Land subsidence was found in the period 1982–1991 in three regions with the highest accumulated subsidence value compared to other regions, namely, two regions in the northwestern part (Cengkareng and Kalideres districts) and the third in the northeastern part of Jakarta (Kemayoran-Sunter district); accumulated subsidence was found in Kalideres district with up to 68.5 cm between 1982 and 1991 with an annual rate of subsidence around 8 cm/year. Accumulated land subsidence was found in the Cengkareng district, with up to 60 cm between 1982 and 1991, with an annual subsidence rate of around 7 cm/year. Accumulated land subsidence was found in the Kemayoran-Sunter district, with up to 70 cm with an annual subsidence rate of around 6 cm/year. The second period of monitoring land subsidence using a leveling survey between 1991 and 1997 highlighted one region with the highest accumulated subsidence value than other regions in the Kalideres district, with up to 154.1 cm with an annual subsidence rate of around 23 cm/year [7].

Following this research, land subsidence in Jakarta was monitored using GPS surveys from 1997 to 2005, and two regions affected by land subsidence were reported in two stations: (1) Kwitang district and (2) Pantai Mutiara district. The accumulated land subsidence was found in the Kwitang district, with up to 48 cm with an annual rate of subsidence rate of 5 cm/year; accumulated land subsidence in Pantai Mutiara district was around 50 cm with an annual rate of subsidence around 4.6 cm/year. [6]. Land subsidence in Jakarta measured using leveling and GPS surveys positively correlated with excessive groundwater extraction and sea-level rise [6,7,49].

Land subsidence was also reported using InSAR techniques based on ALOS PALSAR satellite data in 2007–2009 using the SBAS method, with land subsidence found in Pluit district with an annual subsidence rate of 21.6 cm/year. Land subsidence was found in Cengkareng district with an annual subsidence rate around 21.8 cm/year, in Bekasi district with an annual rate of subsidence of around 10.6 cm/year, and in Karawang district with an annual rate of around 16.4 cm/year [1]. Research was also conducted with ALOS PALSAR satellite data within the period 2007–2010 using the GEOS-PSI method, finding land subsidence in the coastal area and lowland area in northwestern Jakarta within Penjaringan and Cengkareng districts with an annual subsidence rate of up to 26 cm/year, with accumulated subsidence rates of up to 86.5 cm between 2007 and 2010. Land subsidence was observed in the Bekasi district with an annual subsidence rate of up to 11.5 cm/year [46].

3. Material and Methods

3.1. SAR Datasets

A land subsidence inventory map for generating the land subsidence susceptibility map in Jakarta was generated using Sentinel-1 SAR C-band data (5.5 cm wavelength) provided by the European Space Agency (ESA). The SAR data were acquired from March 2017 to May 2020 (91 datasets in the ascending track with path number 98 and frame number 1160, with vertical–vertical (VV) polarization) and in the period of March 2017 to April 2020 (89 datasets in the descending track with path number 47 and frame number 614, with vertical–vertical (VV) polarization). The ascending and descending datasets are listed in Tables 1 and 2, and the reference dates with zero delta day and zero perpendicular baselines from the ascending track are shown on 15 October 2018, whereas those from the descending track are shown on 16 November 2018; both reference dates are shown in bold text in table number 45. A perpendicular baseline graph (generated from Tables 1 and 2 and shown in Figure 2) was used to visualize the temporal baseline from the reference date.

Table 1. Acquisition dates (format ddmmyyyy) of the Sentinel-1 synthetic aperture radar (SAR) datasets in ascending track. Delta days = number of days between each acquisition date. $B\perp$ = perpendicular baseline. The reference dates in ascending tracks are shown in bold text.

No.	Acquisition Date (ddmmyyyy)	DeltaDays	$B_\perp$ (m)	No.	Acquisition Date (ddmmyyyy)	DeltaDays	$B_\perp$ (m)	No.	Acquisition Date (ddmmyyyy)	DeltaDays	$B_\perp$ (m)
1	18032017	−576	77	32	30042018	−168	64	62	07052019	204	93
2	30032017	−564	65	33	12052018	−156	−2	63	19052019	216	16
3	11042017	−552	3	34	24052018	−144	28	64	31052019	228	20
4	23042017	−540	20	35	05062018	−132	127	65	12062019	240	166
5	05052017	−528	−25	36	17062018	−120	103	66	06072019	264	104
6	17052017	−516	17	37	11072018	−96	95	67	18072019	276	44
7	29052017	−504	110	38	23072018	−84	62	68	30072019	288	90
8	10062017	−492	21	39	04082018	−72	98	69	11082019	300	−9
9	22062017	−480	21	40	16082018	−60	75	70	23082019	312	1
10	04072017	−468	112	41	28082018	−48	61	71	04092019	324	46
11	09082017	−432	54	42	09092018	−36	60	72	16092019	336	106
12	21082017	−420	91	43	21092018	−24	55	73	28092019	348	−14
13	02092017	−408	50	44	03102018	−12	115	74	10102019	360	−110
14	14092017	−396	22	45	15102018	0	0	75	22102019	372	−125
15	26092017	−384	43	46	27102018	12	53	76	03112019	384	19
16	08102017	−372	48	47	08112018	24	85	77	15112019	396	−2
17	20102017	−360	72	48	20112018	36	85	78	27112019	408	38
18	01112017	−348	43	49	02122018	48	85	79	09122019	420	98
19	13112017	−336	91	50	14122018	60	142	80	21122019	432	87
20	25112017	−324	30	51	26122018	72	6	81	02012020	444	92
21	07122017	−312	140	52	07012019	84	74	82	14012020	456	40
22	19122017	−300	60	53	19012019	96	46	83	26012020	468	22
23	31122017	−288	135	54	31012019	108	40	84	07022020	480	36
24	12012018	−276	81	55	12022019	120	103	85	19022020	492	74
25	24012018	−264	78	56	24022019	132	29	86	02032020	504	93
26	05022018	−252	37	57	08032019	144	−26	87	14032020	516	31
27	17022018	−240	15	58	20032019	156	26	88	26032020	528	41
28	01032018	−228	30	59	01042019	168	8	89	07042020	540	40
29	13032018	−216	101	60	13042019	180	75	90	19042020	552	14
30	06042018	−192	139	61	25042019	192	−31	91	01052020	564	67
31	18042018	−180	126								

Table 2. Acquisition dates (format ddmmyyy) of the Sentinel-1 Sentinel-1 synthetic aperture radar SAR datasets in descending track. Delta days = number of days between each acquisition date. $B\perp$ = perpendicular baseline. The reference dates in descending tracks are shown in bold text.

No.	Acquisition Date (ddmmyyyy)	DeltaDays	$B_\perp$ (m)	No.	Acquisition Date (ddmmyyyy)	DeltaDays	$B_\perp$ (m)	No.	Acquisition Date (ddmmyyyy)	DeltaDays	$B_\perp$ (m)
1	26032017	−600	22	31	02042018	−228	−20	61	27052019	192	24
2	07042017	−588	76	32	14042018	−216	40	62	08062019	204	0
3	19042017	−576	83	33	26042018	−204	0	63	20062019	216	−7
4	01052017	−564	60	34	08052018	−192	26	64	02072019	228	16
5	13052017	−552	0	35	20052018	−180	0	65	14072019	240	29
6	06062017	−528	0	36	13062018	−156	26	66	07082019	264	87
7	18062017	−516	48	37	25062018	−144	26	67	19082019	276	84
8	30062017	−504	0	38	31072018	−108	−42	68	31082019	288	−5
9	12072017	−492	0	39	24082018	−84	−1	69	12092019	300	0
10	24072017	−480	−10	40	17092018	−60	0	70	24092019	312	117
11	05082017	−468	41	41	29092018	−48	30	71	06102019	324	94
12	17082017	−456	0	42	11102018	−36	51	72	18102019	336	16
13	29082017	−444	57	43	23102018	−24	0	73	30102019	348	19
14	10092017	−432	0	44	04112018	−12	0	74	11112019	360	45
15	22092017	−420	−4	**45**	**16112018**	**0**	0	75	23112019	372	0
16	04102017	−408	0	46	28112018	12	−18	76	05122019	384	5
17	16102017	−396	16	47	10122018	24	3	77	17122019	396	54
18	28102017	−384	96	48	22122018	36	11	78	29122019	408	63
19	09112017	−372	0	49	03012019	48	0	79	10012020	420	35
20	21112017	−360	62	50	15012019	60	24	80	22012020	432	25
21	03122017	−348	−2	51	27012019	72	0	81	03022020	444	11
22	15122017	−336	−23	52	08022019	84	1	82	15022020	456	24
23	27122017	−324	−50	53	20022019	96	34	83	27022020	468	−8
24	08012018	−312	0	54	04032019	108	101	84	10032020	480	91
25	20012018	−300	0	55	16032019	120	23	85	22032020	492	72
26	01022018	−288	14	56	28032019	132	0	86	03042020	504	33
27	13022018	−276	39	57	09042019	144	−12	87	15042020	516	0
28	25022018	−264	0	58	21042019	156	133	88	27042020	528	0
29	09032018	−252	−36	59	03052019	168	82	89	09052020	540	0
30	21032018	−240	−68	60	15052019	180	42				

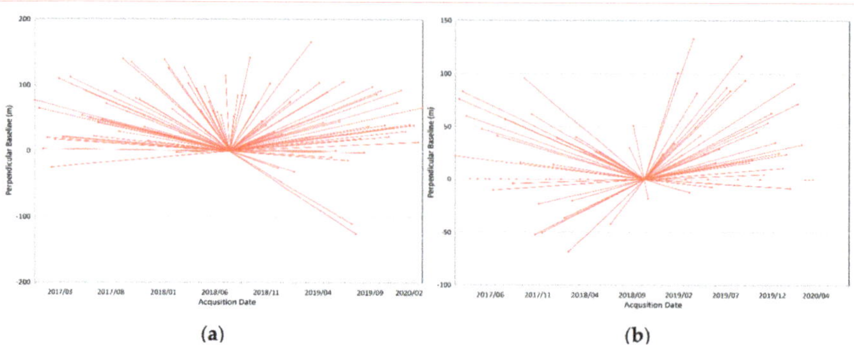

Figure 2. Perpendicular baseline graph from (**a**) ascending track and (**b**) descending track.

3.2. Land Subsidence Conditioning Factors

In total, 14 land subsidence conditioning factors consisting of geological, geomorphological, topographical, and hydrological data were chosen as the conditioning factors (Table 3) in this study as they are considered the main factors influencing land subsidence [11,13,14]. Each factor was classified using the quantile method, and the factors with different cell size resolutions were resampled into raster datasets with 30 m cell size from each conditioning factor to standardize each factor's resolution.

Table 3. Land subsidence conditioning factors in the study area. DEM, digital elevation model; SRTM, Shuttle Radar Topography Mission.

Category	Factor	Source
Hydrological factors	Groundwater drawdown	Groundwater Conservation Center of Indonesia, The Ministry of Energy and Mineral Resources
Hydrological factors	Rainfall intensity	Meteorology, Climatology, and Geophysical Agency of Indonesia
Land cover factors	Road network	Geospatial Information Agency of Indonesia
Hydrological factors	River network	Geospatial Information Agency of Indonesia
Geological factors	Faults	Geospatial Information Agency of Indonesia
Land cover factors	Land use	The Ministry of Environment and Forestry of Indonesia
Geological factors	Lithology	The Ministry of Energy and Mineral Resources
Topographical factors	Elevation	DEM SRTM 1 Arc-Second Global
Topographical factors	Slope	DEM SRTM 1 Arc-Second Global
Topographical factors	Aspect	DEM SRTM 1 Arc-Second Global
Geomorphological factors	Profile curvature	DEM SRTM 1 Arc-Second Global
Geomorphological factors	Plan curvature	DEM SRTM 1 Arc-Second Global
Hydrological factors	Topographic wetness index	DEM SRTM 1 Arc-Second Global

A recent study in Jakarta found that the leading causes of land subsidence in Jakarta are groundwater extraction, the load of buildings and construction, natural consolidation of alluvium soil, and tectonic activities [3]. In this study, groundwater drawdown data were collected from the Groundwater Conservation Center, The Ministry of Energy and Mineral Resources of Indonesia. The observation of groundwater drawdown data is carried out periodically through an automatic water level record (AWLR) system. The annual change in groundwater level from 2019 to 2020 was calculated from 15 monitoring wells, and the map was constructed (Figure 3a) using the inverse distance weighting (IDW) interpolation method from the annual change in groundwater level data. The obtained groundwater drawdown data in this study were insufficient compared to the study area due to the limited monitoring wells over the study area. Although the available data are few, the use of groundwater drawdown data is essential to determine the relationship between land subsidence and groundwater extraction.

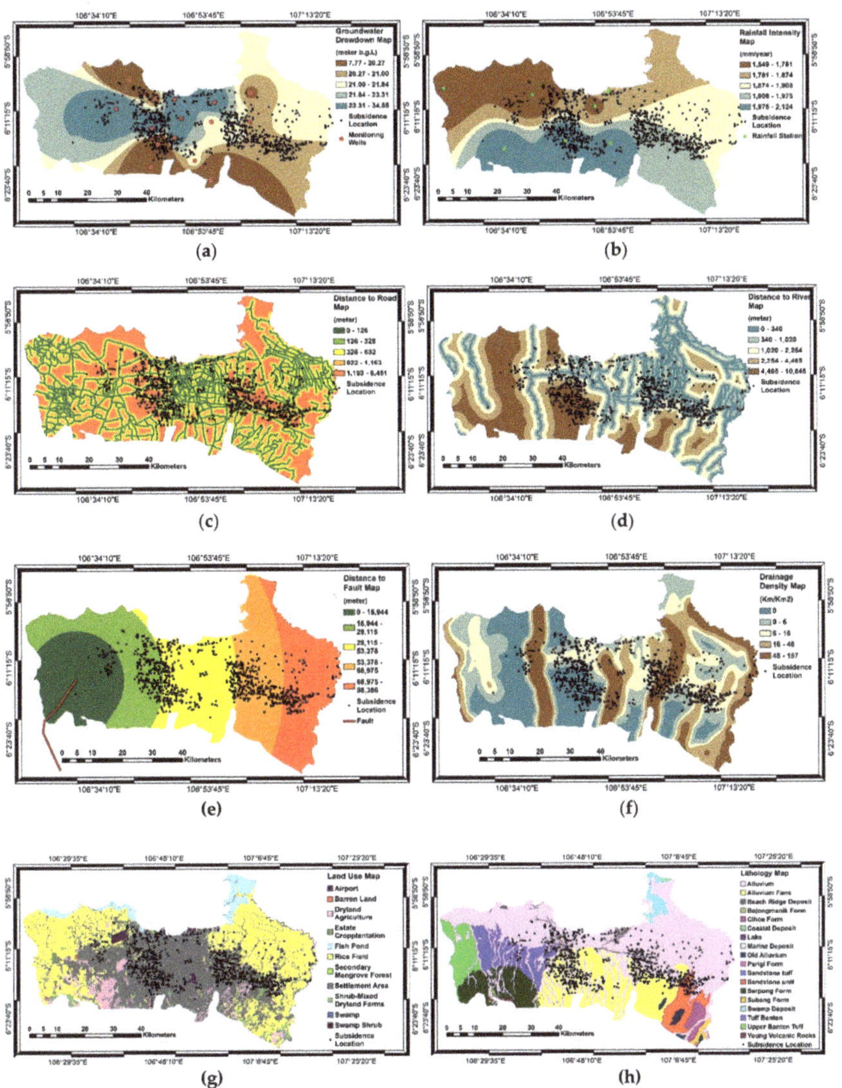

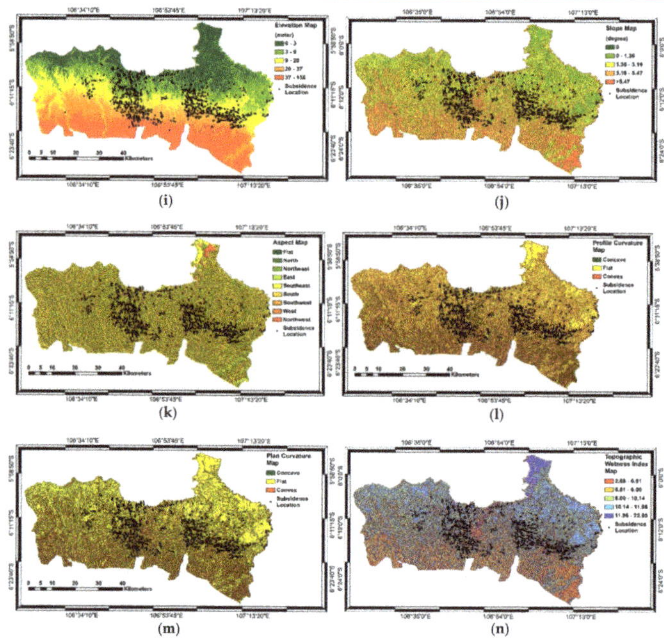

Figure 3. Land subsidence conditioning factors: (**a**) groundwater drawdown; (**b**) rainfall intensity; (**c**) distance to roads; (**d**) distance to rivers; (**e**) distance to faults; (**f**) drainage density; (**g**) land use; (**h**) lithology; (**i**) elevation; (**j**) slope; (**k**) aspect; (**l**) profile curvature; (**m**) plan curvature; (**n**) topographic wetness index.

The groundwater level can increase due to conditional factors indirectly associated with land subsidence, such as rainfall, distance from a river, and river density, which can recharge the groundwater level [13,50]. Four years of daily rainfall data (from 2017 to 2020) from seven weather stations were acquired from the Indonesian Meteorology, Climatology, and Geophysical Agency of Indonesia, and the annual rainfall intensity was calculated and interpolated using the IDW tool in geographic information system (GIS) software (ArcGIS; ESRI, Redlands, CA, USA) (Figure 3b). However, the availability of rainfall data was limited compared to the study area due to the cost and constraints of acquiring data. Nevertheless, the utilization of the rainfall data in this study is important to correlate the topographical factor related to the infiltration flow affecting the soil's strength.

Road, river, and fault networks were acquired from the Geospatial Information Agency of Indonesia from the main road map, main river map, and fault map at a scale of 1:250,000 of polygonal-shaped data using the Atlas map of Indonesia. The information on the location of roads, rivers, and faults was used to construct a distance map (Figure 3c, Figure 3d, and Figure 3e, respectively) using the Euclidean distance tool, and the maps were classified using the quantile method to provide suitable classes within a 30 m cell size. The drainage density or river map density was estimated using the kernel density tool (Figure 3f). Distance to the road and land use are related to urban development in Jakarta, which can affect land subsidence [3]. The land-use map (Figure 3g) was acquired from the Ministry of Environment and Forestry of Indonesia that used Landsat data to generate a land cover map. The geological parameters describe the spatial correlation of lithological landform and land subsidence in Jakarta as being caused by the compaction of the alluvium soil landform [3]. The lithology map (Figure 3h) was acquired from the Geological Atlas map of Indonesia from the Ministry of Energy and Mineral Resources, and the polygonal-shaped map was converted into a raster map with 30 m cell size using GIS tools.

The topographical map, which includes elevation, slope, aspect, profile curvature, plan curvature, and topographic wetness index (TWI) (Figure 3i, Figure 3j, Figure 3k, Figure 3l, Figure 3m, and Figure 3n, respectively) data extracted from the digital elevation model (DEM) of the Shuttle Radar Topography Mission (SRTM) [51], was constructed using the basic terrain analysis tools. Elevation influences the hydrological properties and soil moisture, whereby a lower-elevation area possibly gains more precipitation than a higher-elevation area [13,14]. The slope is associated with land subsidence because it affects the infiltration of rainfall (a steeper surface slope decreases infiltration) [13,21], and the aspect is the second derivative of the slope that has a relationship with land subsidence because the slope aspect affects the strength of the soil due to the moisture preservation and the amount of vegetation [11]. Profile curvature is associated with flow speed, sediment, and erosion quantity, while plan curvature is perpendicular to the slope and indirectly affects land subsidence by influencing convergence and divergence of flow across the surface [13]. The TWI defines the degree of deposition of water at a specific site [52]. The topographical factors extracted from the SRTM DEM are widely used as conditioning factors in land subsidence susceptibility mapping [11–13,28].

The relationship of land subsidence occurrence with the conditioning factors in each class is described in Table 4. A ratio greater than 1 denotes that the class in the conditioning factor is more correlated with the land subsidence occurrence [11]. The calculation of frequency ratio shows that the land subsidence occurred in areas with groundwater level data between 20.27 and 21.00 and between 23.31 and 34.55 m below ground level, while land subsidence also occurred in areas with more annual rainfall intensity due to the recharge of groundwater level and the usage of groundwater. Areas between 0 and 126 m with roads were more correlated with land subsidence occurrence. The land subsidence areas correlated with fault distance were between 15,944 and 68,975 m from the fault location. There were three drainage density classes correlated with land subsidence occurrence and settlement areas correlated with land subsidence. In terms of lithological factors, alluvium, alluvium fans, beach ridge deposits, and sandstone landforms were considered more correlated with the land subsidence occurrences. There were three classes in the elevation map, four classes in the aspect map, two classes in the slope map, one class in the plan curvature, one class in the profile curvature, and three classes in the topographic wetness index map correlated with land subsidence occurrences.

Table 4. Relationship between land subsidence occurrence and conditioning factors using frequency ratio (FR) model.

No.	Conditioning Factor	Class/Category	Ratio each Class	Ratio of Occurrence	FR
1	Groundwater drawdown (m below ground level)	7.77–20.27	0.1831	0.1201	0.6561
		20.27–21.00	0.2021	0.2953	1.4616
		21.00–21.84	0.2361	0.2127	0.9009
		21.84–23.31	0.1914	0.1282	0.6697
		23.31–34.55	0.1874	0.2437	1.3003
2	Rainfall intensity map (mm/year)	1,549–1,781	0.1999	0.1009	0.5046
		1,781–1,874	0.1945	0.1635	0.8404
		1,874–1,908	0.1977	0.2241	1.1338
		1,908–1,975	0.2090	0.3069	1.4680
		1,975–2,124	0.1989	0.2047	1.0290
3	Distance to road map (m)	0–126	0.2114	0.2201	1.0412
		126–328	0.1978	0.1964	0.9931
		328–632	0.1972	0.1943	0.9853
		632–1,163	0.1968	0.1964	0.9979
		1,163–6,451	0.1968	0.1928	0.9795

Table 4. *Cont.*

No.	Conditioning Factor	Class/Category	Ratio each Class	Ratio of Occurrence	FR
4	Distance to river map (m)	0–340	0.2009	0.2262	1.1262
		340–1,020	0.1999	0.2369	1.1848
		1,020–2,254	0.1998	0.2030	1.0159
		2,254–4,465	0.1997	0.1422	0.7120
		4,465–10,845	0.1997	0.1917	0.9601
5	Distance to fault map (m)	0–15,944	0.2000	0.0386	0.1929
		15,944–29,115	0.2000	0.2029	1.0145
		29,155–53,378	0.2000	0.3375	1.6874
		53,378–68,975	0.2000	0.2845	1.4223
		68,975–88,386	0.2000	0.1366	0.6829
6	Drainage density (km/km^2)	0	0.2331	0.2408	1.0331
		0–6	0.1917	0.1655	0.8632
		6–16	0.1917	0.2071	1.0801
		16–48	0.1917	0.2101	1.0960
		48–157	0.1917	0.1765	0.9206
7	Land-use map	Airport	0.0084	0.0022	0.2644
		Barren land	0.0009	0.0004	0.4666
		Dryland agriculture	0.0594	0.0076	0.1275
		Estate crop plantation	0.0035	0.0003	0.0739
		Fish pond	0.0475	0.0012	0.0250
		Rice field	0.3979	0.0650	0.1634
		Secondary mangrove forest	0.0003	0.0001	0.3295
		Settlement area	0.4572	0.9194	2.0110
		Shrub-mixed dryland farms	0.0198	0.0028	0.1440
		Swamp	0.0051	0.0009	0.1831
		Swamp shrub	0.0000	0.0000	0.1441
8	Lithology map	Alluvium	0.5020	0.5577	1.1111
		Alluvium fans	0.1966	0.3200	1.6274
		Beach ridge deposits	0.0153	0.0341	2.2385
		Bojongmanik form	0.0022	0.0000	0.0000
		Cihoe form	0.0161	0.0002	0.0149
		Coastal deposit	0.0013	0.0000	0.0000
		Lake	0.0001	0.0000	0.1178
		Marine deposits	0.0052	0.0000	0.0024
		Old alluvium	0.0074	0.0002	0.0239
		Parigi form	0.0005	0.0000	0.0000
		Sandstone tuff	0.0001	0.0000	0.0000
		Sandstone unit	0.0405	0.0440	1.0873
		Serpong form	0.0703	0.0014	0.0198

Table 4. *Cont.*

		Subang form	0.0064	0.0000	0.0030
		Swamp deposits	0.0175	0.0005	0.0300
		Tuff banten	0.0677	0.0391	0.5779
		Upper banten tuff	0.0488	0.0026	0.0543
		Young volcanic rocks	0.0022	0.0000	0.0000
9	Elevation map (m)	0–3	0.2227	0.1088	0.4883
		3–9	0.2029	0.2170	1.0699
		9–20	0.1976	0.3767	1.9061
		20–37	0.1968	0.2637	1.3397
		37–156	0.1800	0.0338	0.1878
10	Slope (degree)	0	0.2011	0.1727	0.8587
		0–1.36	0.1997	0.1657	0.8294
		1.36–3.19	0.1997	0.2399	1.2009
		3.19–5.47	0.1997	0.2470	1.2367
		>5.47	0.1997	0.1748	0.8753
11	Aspect	Flat	0.1214	0.1339	1.1030
		North	0.1490	0.1778	1.1937
		Northeast	0.1091	0.1323	1.2128
		East	0.1177	0.1232	1.0468
		Southeast	0.1006	0.0888	0.8829
		South	0.1212	0.0924	0.7626
		Southwest	0.0937	0.0711	0.7592
		West	0.0937	0.0880	0.9386
		Northwest	0.0937	0.0925	0.9869
12	Profile curvature	Concave	0.3332	0.3253	0.9764
		Flat	0.3336	0.3017	0.9045
		Convex	0.3332	0.3729	1.1192
13	Plan curvature	Concave	0.3332	0.3253	0.9764
		Flat	0.3336	0.3017	0.9045
		Convex	0.3332	0.3729	1.1192
14	Topographic wetness index	2.52–6.81	0.1430	0.1533	1.0722
		6.81–8.00	0.1939	0.2252	1.1614
		8.00–10.14	0.2140	0.2279	1.0647
		10.14–11.96	0.2258	0.2098	0.9293
		11.96–22.90	0.2233	0.1839	0.8232

3.3. Illustration of Methodology

The methods to generate land subsidence susceptibility maps using functional and meta-ensemble algorithms are described below and illustrated in Figure 4.

1. Land subsidence occurrences were identified by exploiting Sentinel-1 SAR datasets from 2017 to 2020 from both ascending and descending tracks using time-series InSAR techniques based on the StaMPS algorithm. The persistent scatterer points from co-registered single master images showing a deformation value were used as the land subsidence inventory map.
2. Preparation of training and testing datasets was conducted by randomly dividing the persistent scatterer (PS) points of time-series InSAR showing a vertical deformation into 50% training data to generate land subsidence susceptibility models and 50% testing data to validate the land subsidence susceptibility map, as done in other studies finding optimal results [28,53]. The distribution of training and test data is shown in Figure 1b.
3. Preparation of land subsidence conditioning factors for spatial correlation analysis was done using the frequency ratio method to find the correlation between each factor and land subsidence occurrence [53]. We used each model's ratio value and then used as the conditioning factors related to land subsidence occurrences. First, the conditioning factors were classified using

quantile methods in GIS tools with a similar environment of 30 m cell size for each factor; then, the number of subsidence occurrences in each class was calculated using the cross-tabulation tool in GIS. Next, we calculated the ratio between the percentage of pixels of each conditioning factor class and the percentage of subsidence occurrence pixels to obtain the FR value as follows:

$$\text{Frequency Ratio} = \frac{\% \text{ of class conditioning factor}}{\% \text{ of land subsidence occurrence}} \tag{1}$$

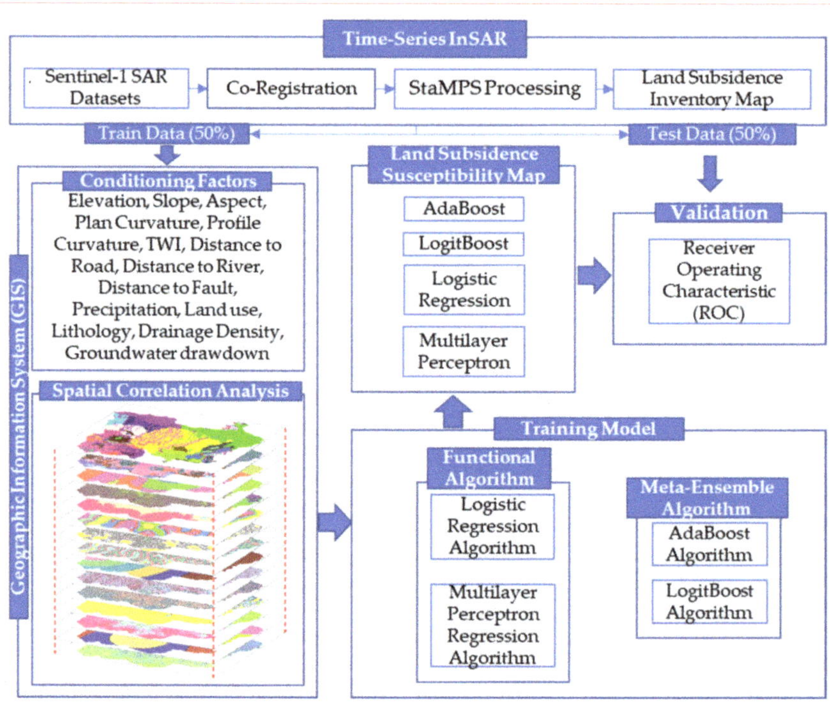

Figure 4. The study workflow.

4. The conditioning factors consisting of frequency ratio values were used to generate land subsidence susceptibility models using two functional algorithms (logistic regression and multilayer perceptron) and two meta-ensemble algorithms (AdaBoost and LogitBoost).

5. After all land subsidence susceptibility maps were generated, all maps were validated using the test data prepared before and analyzed using ROC curve analysis.

3.4. StaMPS Processing

StaMPS (Stanford Method for Persistent Scatterers) is an analysis method used to facilitate the generation of time-series deformation images of all terrains, including nonurban areas. The StaMPS algorithm uses the spatial correlation of phase measurements rather than a functional temporal model to identify PS pixels. StaMPS processing does not use a model to describe how the displacement rate varies with time. To identify PS pixels in a single master stack of interferograms, StaMPS uses the phase characteristics from the dominant point scatterer in each area and creates interferograms from SAR images. It also reduces decorrelation [54]. Thus, the StaMPS algorithm can identify and extract the deformation signal from stable pixels in all terrains.

The slave images in the acquired SAR datasets were resampled to perfectly match the master image through a co-registration process before generating an interferogram. The co-registration process was applied to two different images to produce refined SAR images, which were then cropped to focus only on the area of interest. Next, differential InSAR (DInSAR) images were generated by subtracting the topographic InSAR images generated using the interferograms from the topographic phases of the SRTM DEM [55,56]. We used the PS method to measure the displacement of the earth's surface [57] to generate the time-series deformation map. The main processes generated a single master stack of interferograms and topographic phase removal [33].

The StaMPS algorithm is shown in Figure 5. Multiple images were co-registered to generate a single master image for the ascending track on 15 October 2018 and the descending track on 16 November 2018; co-registered images with the topographic phases removed were subjected to amplitude and phase noise analysis to derive a subset comprising all PS pixels, with weeding performed using a threshold value of 0.4. The wrapped phase of the selected pixels was corrected for the spatially uncorrelated look-angle error. The phase was then unwrapped, and PS outputs were generated. The deformation map from the line of sight (LOS) displacement could be converted into vertical deformation data [58,59] by assuming the horizontal deformation as very small compared to the vertical deformation caused by land subsidence [60–62]. Recent studies using GPS and leveling surveys reported that the land subsidence in Jakarta shows a vertical deformation [7], and a vertical deformation pattern was also found in research using InSAR to monitor land subsidence in Jakarta [1,46]; hence, the deformation from the line of sight (LOS) in this study could be assumed as negligible and could be converted directly into vertical deformation value using Equation (2) by dividing the displacement or deformation from the line of sight (d_{LOS}) by the cosine of incident angle (θ) from the radar signal. The results of vertical deformation were assigned a negative value from the initial ground-level observation point, indicating that the land subsidence occurred vertically at that point [9].

$$V = \frac{d_{LOS}}{\cos \theta} \tag{2}$$

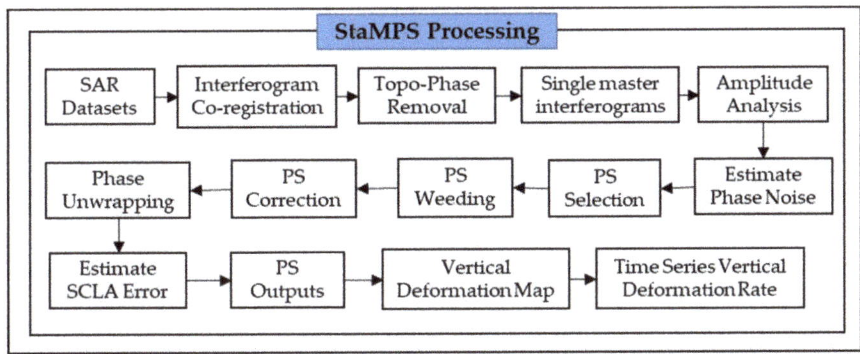

Figure 5. Flowchart of Stanford Method for Persistent Scatterers (StaMPS) Processing.

3.5. AdaBoost

AdaBoost is a machine learning algorithm introduced by Freund and Schapire (1997) [63]. AdaBoost's classifier uses an adaptive resampling technique that produces a series of individual classifiers to classify training samples accurately. The frequency of variables selected by a weak learner was examined, and the relative importance of the variables could be determined [64]. AdaBoost combines multiple weak learners to derive a single strong learner by repeatedly calling a weak classifier and adjusting the attributed weight to the sample. The new classifier focuses more on the misclassified sample as the misclassified sample is increased compared to the correct sample. The final weight is obtained by adding or subtracting the updated weight in each iteration. The final model is obtained by dividing each final weight by the total adjusted weight [65]. The general form of the AdaBoost algorithm is as follows [66]:

1. Start with weights $w_i = \frac{1}{N}$ for $i = 1, \ldots, N$;
2. Repeat this step for $m = 1, \ldots, M$:

   a. Fit the classifier $f_m(x) \in \{-1, 1\}$ using weights w_i with the training data;

   b. Compute $err_m = E_w\left[1_{(y \neq f_m(x))}\right]$, $c_m = \log\left(\frac{1 - err_m}{err_m}\right)$;

   c. Set $w_i \leftarrow w_i \exp\left[c_m 1_{(y \neq f_m(x))}\right]$, $i = 1, 2, \ldots, N$, and renormalize so that $\sum_i w_i = 1$;

3. Output the classifier: $\text{sign}\left[\sum_{m=1}^{M} c_m f_m(x)\right]$.

3.6. LogitBoost

LogitBoost is a modified version of the AdaBoost algorithm, introduced by Friedman, Hastie, and Tibshirani [66], where the exponential loss function is replaced with the log-likelihood loss function. This method reduces classification error and is less sensitive to noise [28,67]. LogitBoost can handle multiple class problems and uses a regression scheme as the base learner for classification [64]. The general form of the LogitBoost algorithm is as follows [66]:

1. Start with weights $w_i = \frac{1}{N}$ for $i = 1, 2, \ldots, N$, $F(x) = 0$, and probability estimates $p(x_i) = \frac{1}{2}$;
2. Repeat this step for $m = 1, \ldots, M$:

   a. Compute the working response and weights:

$$l_i = \frac{y_i^* - p(x_i)}{p(x_i)\left(1 - p(x_i)\right)},\tag{3}$$

$$w_i = p(x_i)\left(1 - p(x_i)\right);\tag{4}$$

   b. Fit the function by weighted least-squares regression of l_i to x_i using weight w_i;

   c. Update the function as follows:

$$f(x) \leftarrow f(x) + \frac{1}{2}f_m(x),\tag{5}$$

$$p(x) \leftarrow \frac{e^{f(x)}}{e^{f(x)} + e^{-f(x)}};\tag{6}$$

3. Output the classifier: $\text{sign}[F(x)] = \text{sign}\left[\sum_{m=1}^{M} f_m(x)\right]$.

3.7. Logistic Regression

Logistic regression is a statistical method used to find the best model to describe the correlation between a dependent variable and several independent variables. This method's advantage is that the variables do not need to be normally distributed [68]. It also offers several ways of selecting the best predictor for use in the P probability model [69,70]. The equations describing the logistic regression are as follows [28,69–71]:

$$f(x) = \text{logit}(P) = \ln\left[\frac{P}{1 - P}\right] = c_0 + c_1 x_1 + \ldots + c_n x_n,\tag{7}$$

$$P = \frac{1}{1 + e^{-f(x)}} = \frac{1}{1 + e^{-(c_0 + c_1 x_1 + \dots + c_n x_n)}}, \tag{8}$$

where $f(x)$ is a linear combination function called $\mathrm{logit}(P)$, P is the probability of subsidence occurrence, $1 - P$ is the probability that subsidence will not occur, $x_1, x_2, \dots, x_n$ are input variables, c_0 is the model intercept, and $c_1, \dots, c_n$ are the approximated coefficients of regression.

3.8. Multilayer Perceptron

The multilayer perceptron is a machine learning algorithm based on the ANN technique that consists of input layers, hidden layers, and output layers [72]. The advantages of the multilayer perceptron algorithm are as follows: the distribution of the training data does not require any assumptions, most inputs are selected during the training process based on the weight adjustment, and the relative importance of the various input measures does not need to be determined [73]. The equation representing the multilayer perceptron for land subsidence classification is as follows [28,72,73]:

$$m = f(c), \tag{9}$$

where $f(c)$ is a hidden function that is optimized during the training process for a given network architecture via the adjustable weights, and $c = c_i$ for $i = 1, \dots, 14$, which is a vector containing 14 land subsidence conditioning factors.

4. Results

4.1. Land Subsidence Inventory Map

The mean vertical deformation maps for Jakarta shown in Figure 6a,b were derived by time-series InSAR analysis based on the StaMPS algorithm. We overlaid a true color red/green/blue (RGB) composite image from Sentinel-2 taken on 28 August 2020. Six areas from both ascending and descending tracks showed high deformation; thus, we plotted a deformation trend for these areas.

The vertical deformation rate in the ascending track at Point P1 (Figure 6c) represents the Kosambi area, which subsided 189.48 mm from 2017 to 2020. Point P2 (Figure 6c) represents the Cengkareng area, which subsided 184.02 mm from 2017 to 2020. Point P3 (Figure 6e) represents the Ciledug area, which subsided up to 155.22 mm from 2017 to 2020. Point P4 (Figure 6e) represents the Penjaringan area, which subsided 148.36 mm from 2017 to 2020. Point P5 (Figure 6g) represents the Bekasi area, which subsided 128.17 mm from 2017 to 2020. Point P6 (Figure 6g) represents the Cikarang area, which subsided 271.84 mm from 2017 to 2020.

The vertical deformation rate in the descending track at Point P1 (Figure 6d) represents the Kosambi area, which subsided 210.07 mm from 2017 to 2020. Point P2 (Figure 6d) represents the Cengkareng area, which subsided 216.19 mm from 2017 to 2020. Point P3 (Figure 6f) represents the Ciledug area, which subsided up to 155.92 mm from 2017 to 2020. Point P4 (Figure 6f) represents the Penjaringan area, which subsided 148.31 mm from 2017 to 2020. Point P5 (Figure 6h) represents the Bekasi area, which subsided 107.19 mm from 2017 to 2020. Point P6 (Figure 6h) represents the Cikarang area, which subsided 257.94 mm from 2017 to 2020.

Figure 6g,h show that the deformation pattern was mostly linear. However, Figure 6c–f, representing the Kosambi, Cengkareng, Ciledug, and Penjaringan areas, show quite periodic subsidence with the standard deviation of the vertical deformation rate being higher than the vertical deformation rate in Figure 6g,h. These results occurred due to the seasonal effect of groundwater extraction, and those areas were surrounded by a residential area that mostly used groundwater as the water source. Meanwhile, P6 is one of the most significant industrial areas in Indonesia. The mean vertical deformation maps from ascending and descending tracks were compared to validate the accuracy of the land subsidence inventory map using the StaMPS algorithm, and the result showed a good correlation in Figure 6i with a coefficient of correlation (R^2) up to 0.9584 between ascending and descending tracks.

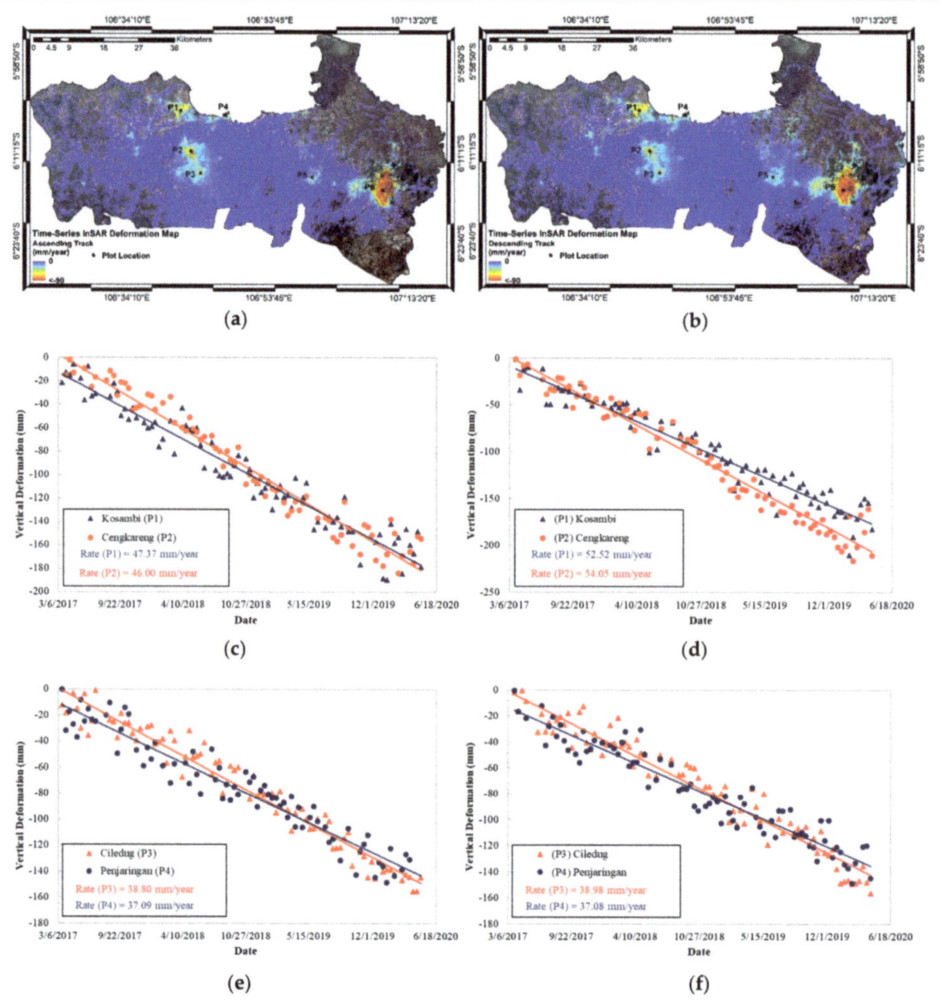
106°34'10"E
106°53'45"E
107°13'20"E
Kilometers
0 4.5 9 18 27 36
Time-Series InSAR Deformation Map
Ascending Track
(mm/year)
Plot Location
0
<-90
(a)
106°34'10"E
106°53'45"E
107°13'20"E
Time-Series InSAR Deformation Map
Descending Track
(mm/year)
Plot Location
0
<-90
(b)
Vertical Deformation (mm)
Kosambi (P1)
Cengkareng (P2)
Rate (P1) = 47.37 mm/year
Rate (P2) = 46.00 mm/year
Date
(c)
(P1) Kosambi
(P2) Cengkareng
Rate (P1) = 52.52 mm/year
Rate (P2) = 54.05 mm/year
Date
(d)
Ciledug (P3)
Penjaringan (P4)
Rate (P3) = 38.80 mm/year
Rate (P4) = 37.09 mm/year
Date
(e)
(P3) Ciledug
(P4) Penjaringan
Rate (P3) = 38.98 mm/year
Rate (P4) = 37.08 mm/year
Date
(f)

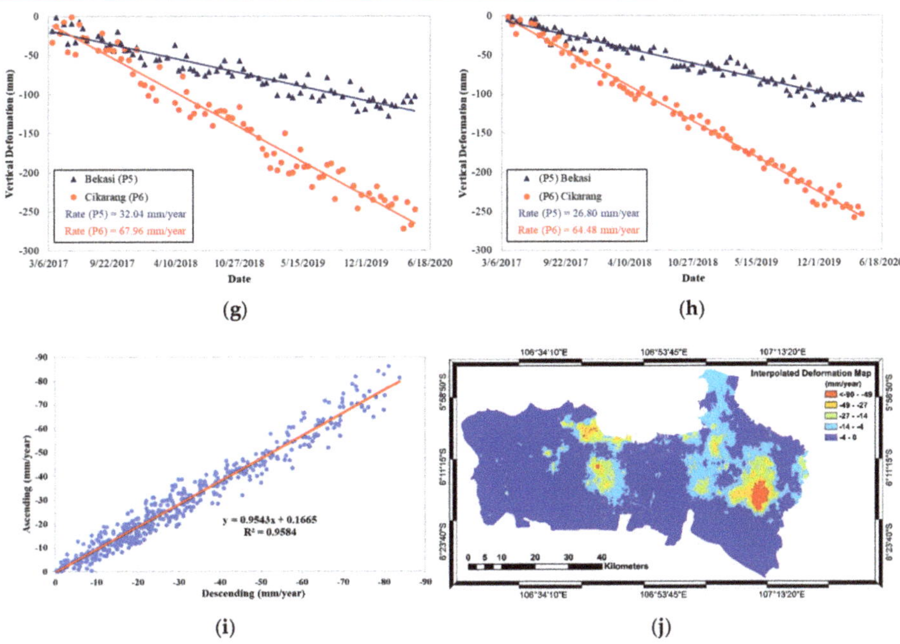

Figure 6. Mean vertical deformation map for Jakarta depicted using Sentinel-2 image in (**a**) ascending tracks and (**b**) descending tracks; the vertical deformation rate at P1 (Kosambi) and P2 (Cengkareng) in (**c**) ascending and (**d**) descending tracks; the vertical deformation rate at P3 (Ciledug) and P4 (Penjaringan) in ascending (**e**) and descending (**f**) tracks; the vertical deformation at P5 (Bekasi) and P4 (Cikarang) in ascending (**g**) and descending tracks (**h**). (**i**) The comparison of mean vertical deformation between ascending and descending tracks. (**j**) Kriging interpolation of time-series deformation map from mean vertical deformation map of descending track, resulting in the extension of the land subsidence inventory map; the blue area is considered as the nonoccurrence area of land subsidence.

The persistent scatterer density of both tracks in the east and west areas was relatively low due to them being wetland areas and more vegetated than other areas. The SAR dataset used in this study from Sentinel-1 SAR C-band data with 5.5 cm wavelength could not deeply penetrate beneath the trees. Thus, to overcome that limitation, the interpolation of the persistent scatterer points from the descending track was constructed using kriging interpolation in GIS tools to provide land subsidence information over the study area as shown in Figure 6j [74].

4.2. Land Subsidence Susceptibility Map

The land subsidence susceptibility model's performance depended on the calculated parameters (Table 5) for optimization used in this study.

A land subsidence susceptibility map was generated using 14 land subsidence conditioning factors, training data from our land subsidence inventory map, and four different algorithms: LogitBoost (Figure 7a), AdaBoost (Figure 7b), logistic regression (Figure 7c), and multilayer perceptron (Figure 7d). Land subsidence susceptibility indices were generated for all unique pixels in the study area. The susceptibility indices were reclassified using the quantile method to identify feature pairs in five susceptibility classes: very low, low, moderate, high, and very high [28,64].

Table 5. Calculated parameters for the algorithms used in this study.

Algorithm	Parameters
AdaBoost	The number of iterations: 10; seed: 1; weight threshold: 100.
LogitBoost	Number of iterations: 10; Seed: 1; weight threshold: 100; likelihood threshold: -1.7976E308; shrinkage: 1.0; max threshold: 3; thread pool: 1; thread to batch prediction: 1.
Logistic Regression	Ridge: 1.0E-8; max iterations: -1; number of decimal places: 4.
Multilayer Perceptron	Hidden layers: a; learning rate: 0.3; momentum: 0.2; number of decimal places: 2; seed: 0; training time: 500; validation set size: 0; validation threshold: 20.

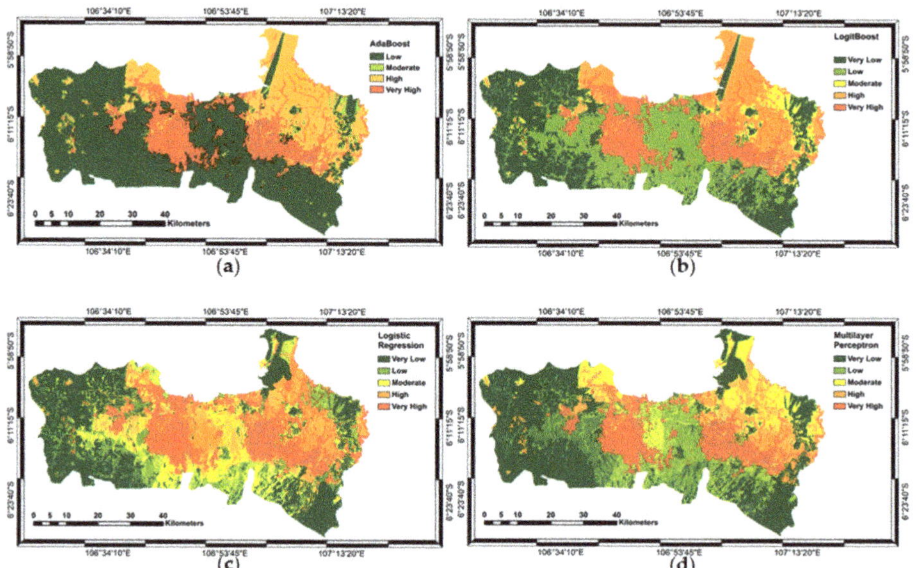

Figure 7. Land subsidence susceptibility map for Jakarta, generated using four different algorithms: (**a**) AdaBoost, (**b**) LogitBoost, (**c**) logistic regression, and (**d**) multilayer perceptron.

The proportion of very high susceptibility land was quite similar for all four algorithms. However, the map generated by AdaBoost differed from those of the other three methods because the AdaBoost model could only classify susceptibility into four classes due to the limits of the probability range. Therefore, the very low susceptibility class was excluded. The susceptibility class proportions (pixel distributions) are shown in Figure 8. For the Adaboost algorithm, the proportions were 0%, 57.88%, 2.54%, 22.02%, and 17.56% for the very low, low, moderate, high, and very high classes, respectively; the respective values for the LogitBoost algorithm were 33.33%, 27.09%, 7.03%, 16.77%, and 15.78%, and those for the logistic regression algorithm were 32.62%, 13.86%, 16.44%, 18.51%, and 18.57%. Finally, the multilayer perceptron algorithm's respective values were 40.64%, 18.70%, 14.02%, 13.42%, and 13.23%.

The distribution of pixels in the very high class and high class in Figure 8 showed similar results for each algorithm, resulting in the maps of these classes being quite similar. The very high class was considered to be the land subsidence areas shown in the mean vertical deformation map in Figure 6a,b, with similarity seen because of the training data used in this study being acquired from the land subsidence inventory map with a large spatial resolution, which allowed more effectively defining the land subsidence area in the susceptibility map. The moderate and high classes were considered land subsidence areas in the future, and the very low and low classes were areas with the lowest probability of land subsidence in the future. The difference between the moderate class from AdaBoost and that from other algorithms is that AdaBoost did not consider other areas far from the land subsidence occurrences. Other algorithms showed that residential areas located in alluvium landforms had a reasonable possibility of land subsidence.

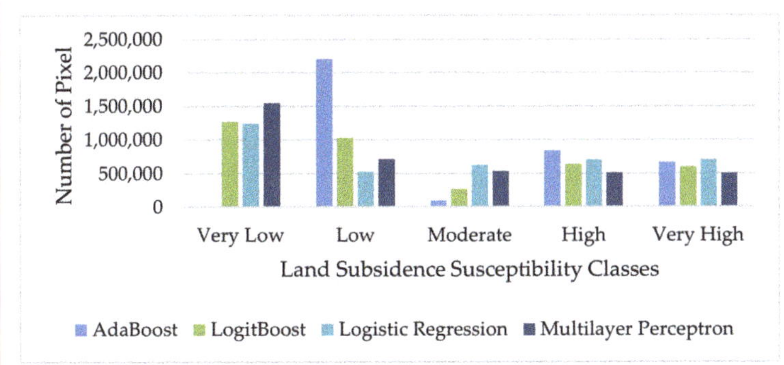

Figure 8. Proportions of susceptibility classes for land subsidence susceptibility maps generated using four different machine learning algorithms.

4.3. Model Validation

The accuracy of all four algorithms in this study was evaluated by ROC curve analysis [12,17]. ROC curve analysis is a standard way of validating the probability models used to generate land subsidence susceptibility maps, according to the area under the curve (AUC) [22,28]. Higher values indicate more accurate and reliable models. If the AUC, which ranges from 0 to 1, is lower than 0.5, the model is considered unacceptably inaccurate [75].

Land subsidence susceptibility maps produced using functional (AdaBoost, LogitBoost) and meta-ensemble (logistic regression and multilayer perceptron) algorithms were compared. The ROC curves for the four algorithms are shown in Figure 9. The largest AUC of 0.811 was from the AdaBoost algorithm (blue line in Figure 9). The multilayer perceptron algorithm had the next largest AUC (0.800; purple line in Figure 9), followed by the logistic regression (0.794; green line in Figure 9) and LogitBoost algorithms (0.791; red line in Figure 9).

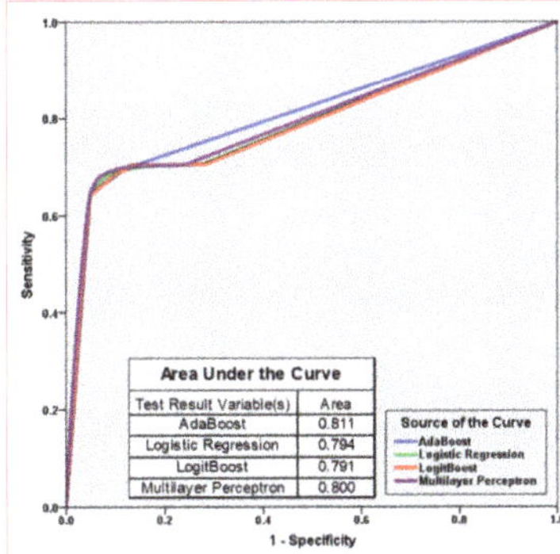

Figure 9. Receiver operating characteristic (ROC) curves for the land subsidence susceptibility maps produced by functional (logistic regression and multilayer perceptron) and meta-ensemble (AdaBoost and LogitBoost) algorithms.

Variables having at least one tie between the positive actual state group and the negative actual state group can bias the results. Since all of the AUC values were higher than 0.5, the land subsidence susceptibility maps produced by all algorithms used in this study are acceptable for predicting high-risk subsidence areas in Jakarta [75]. However, the AdaBoost algorithm map had the best performance.

5. Discussion

5.1. Land Subsidence Inventory Map

A land subsidence inventory map was successfully created through time-series InSAR analysis of Sentinel-1 datasets, using the StaMPS algorithm. The map is displayed as a vertical deformation map in Figure 6a,b for descending and ascending tracks. The comparison was made to validate the mean vertical deformation's accuracy used as a land subsidence inventory map. A coefficient of correlation of 0.9584 was found. As stated above, there were six high-deformation areas (Figure 6c–h) within ascending and descending tracks, which were equally affected (in terms of subsidence) by increased groundwater usage and existing buildings [3].

Figure 6g,h show areas with largely linear deformation rates, the standard deviations of which were smaller than those of the area shown in Figure 6c–f. The periodic subsidence shown was affected due to the variable climate in Jakarta, which is generally characterized by high rainfall but a dry summer. In the rainy season, the groundwater level of all aquifers beneath Jakarta increases, which manifests as deformation, i.e., an uplift in the ground level due to a change in the underlying material's thickness [34,76–79].

As shown by the land subsidence occurrence (Figure 6a,b) and groundwater drawdown (Figure 3a) data, land subsidence for P1–P5, which are between 20.58 and 34.55 m below ground level, correlated with the groundwater level. The correlation between land subsidence and groundwater level might have been stronger if groundwater level data were available for the whole study area. The groundwater level is deepest for P1 and P2, ranging from 22.47 to 34.55 m. P6 has more significant deformation because it contains the largest industrial estate in Indonesia.

Jakarta is mostly situated on young alluvial soil (Figure 3h), which cannot tolerate the maximum compression force of many buildings [9]. Thus, compaction of the unconsolidated alluvial soil occurs and is exacerbated by groundwater extraction (because pore pressure is reduced, leading to further clay compaction) [80,81].

5.2. Land Subsidence Susceptibility Map

The method used to create land subsidence susceptibility maps strongly affects the quality of the mapping. Machine learning techniques are effective [12,13,28,64]. In particular, the method used to generate training and testing data is important. Accurate land subsidence inventory maps can be obtained using InSAR; we combined InSAR and GIS spatial data to produce an accurate land subsidence susceptibility map [12]. Nevertheless, the distribution data of groundwater drawdown and rainfall intensity data from the interpolation process might not represent the whole study area. They could have affected the spatial distribution of the raster map. Access to the monitoring wells and rainfall station outside the Jakarta area could provide more accurate analysis to determine the relationship between those factors and land subsidence.

The land subsidence susceptibility maps of all four algorithms used in this study could be ordered according to the accuracy and time taken to build the model due to the similarities in the conditioning factors, training and testing data, and study area [28]. We used ROC curve analysis to assess the accuracy of the maps. The AUC data showed that the AdaBoost algorithm had the highest susceptibility class predictive accuracy of 81.1%, which was 1.1% higher than the multilayer perceptron algorithm, 1.7% higher than the logistic regression algorithm, and 2% higher than the LogitBoost algorithm. Since the accuracy of the algorithms used was so closely related, we analyzed the time consumption of data preparation needed to build the model. The AdaBoost algorithm needed 218.63 s to build the model, the LogitBoost algorithm needed 174.94 s to build the model, logistic regression needed 16667.96 s or 4.63 h to build the model, and multilayer perceptron needed 47,538.65 s or 13.20 h to build the model, thus identifying the LogitBoost algorithm as the fastest.

However, the LogitBoost algorithm had the lowest prediction accuracy (79.10%). The Adaboost algorithm needed 218.63 s or 3.64 min to build the model, along with having the highest prediction accuracy of 81.10% in the ROC curve analysis. Logistic regression needed 4.63 h to build the model, along with a prediction accuracy of about 79.40%, and the multilayer perceptron needed 13.20 h, along with a prediction accuracy of 80%. The maps generated using LogitBoost, logistic regression, and multilayer perceptron (Figure 7b–d, respectively) showed similar weights and indicated that the settlement area was at a high risk of land subsidence. The land-use map generated in this study indicated a relatively strong correlation between land subsidence and urban development, likely due to excessive groundwater extraction in urban areas [3,6,7]. Our map could be used by local environmental authorities and those in charge of urban development to identify areas with high subsidence risk. The method of combining SAR and GIS spatial data to generate land subsidence susceptibility maps employed in this study could be applied to other regions.

6. Conclusions

We generated a land subsidence inventory map using a time-series InSAR method based on the StaMPS algorithm from Sentinel-1 SAR datasets in ascending and descending tracks. The comparison of both tracks was conducted, finding a coefficient of correlation between the two tracks of 0.9548. The inventory map could be used as a training and testing dataset to create a land subsidence susceptibility map for Jakarta using meta-ensemble (AdaBoost and LogitBoost) and functional (logistic regression and multilayer perceptron) machine learning algorithms. We created a land subsidence inventory map through time-series InSAR analysis of Sentinel-1 SAR datasets using the StaMPS algorithm. The land subsidence susceptibility map produced by the AdaBoost machine learning algorithm had higher accuracy (AUC = 0.811) and only need 3.64 min to build the model compared with the maps created using the other algorithms (multilayer perceptron, AUC = 0.800; logistic regression, AUC = 0.794; LogitBoost, AUC = 0.791). LogitBoost was the fastest algorithm in building the model but had the lowest predictive accuracy. Logistic regression and multilayer perceptron needed 4.63 and 13.20 h, respectively, to build the model. All of the maps showed acceptable accuracy, as the AUC values were all higher than 0.5; thus, they can all be used for analyzing land subsidence susceptibility in Jakarta. Our approach based on time-series InSAR analysis, machine learning, and GIS spatial data yielded reasonable predictions of areas with high risk of land subsidence. Further research could use alternative algorithms and conditioning factors to generate land subsidence susceptibility maps in other regions.

Author Contributions: Conceptualization, C.-W.L., and W.L.H.; methodology, C.-W.L., A.R.A., and W.L.H.; software, A.R.A. and W.L.H.; validation, C.-W.L. and W.L.H.; formal analysis, W.L.H.; investigation, C.-W.L.; resources, C.-W.L.; data curation, C.-W.L. and W.L.H.; writing—original draft preparation, W.L.H.; writing—review and editing, C.-W.L., A.R.A., and W.L.H.; visualization, C.-W.L., A.R.A., and W.L.H.; supervision, C.-W.L.; project administration, C.-W.L.; funding acquisition, C.-W.L. All authors read and agreed to the published version of the manuscript.

Funding: This research was supported by a grant from the National Research Foundation of Korea provided by the government of Korea (No. 2019R1A2C1085686).

Conflicts of Interest: The authors declare no conflict of interest. The funders had no role in the design of the study; in the collection, analyses, or interpretation of data; in the writing of the manuscript, or in the decision to publish the results.

References

1. Chaussard, E.; Amelung, F.; Abidin, H.; Hong, S.-H. Sinking cities in Indonesia: ALOS PALSAR detects rapid subsidence due to groundwater and gas extraction. *Remote Sens. Environ.* **2013**, *128*, 150–161. [CrossRef]
2. Machowski, R.; Rzetala, M.A.; Rzetala, M.; Solarski, M. Geomorphological and Hydrological Effects of Subsidence and Land use Change in Industrial and Urban Areas. *Land Degrad. Dev.* **2016**, *27*, 1740–1752. [CrossRef]
3. Abidin, H.Z.; Andreas, H.; Gumilar, I.; Fukuda, Y.; Pohan, Y.E.; Deguchi, T. Land subsidence of Jakarta (Indonesia) and its relation with urban development. *Nat. Hazards* **2011**, *59*, 1753–1771. [CrossRef]
4. Takagi, H.; Esteban, M.; Mikami, T.; Fujii, D. Projection of coastal floods in 2050 Jakarta. *Urban Clim.* **2016**, *17*, 135–145. [CrossRef]
5. Budiyono, Y.; Aerts, J.C.J.H.; Tollenaar, D.; Ward, P.J. River flood risk in Jakarta under scenarios of future change. *Nat. Hazards Earth Syst. Sci.* **2016**, *16*, 757–774. [CrossRef]
6. Abidin, H.Z.; Andreas, H.; Djaja, R.; Darmawan, D.; Gamal, M. Land subsidence characteristics of Jakarta between 1997 and 2005, as estimated using GPS surveys. *GPS Solut.* **2008**, *12*, 23–32. [CrossRef]
7. Abidin, H.Z.; Djaja, R.; Darmawan, D.; Hadi, S.; Akbar, A.; Rajiyowiryono, H.; Sudibyo, Y.; Meilano, I.; Kasuma, M.A.; Kahar, J.; et al. Land subsidence of Jakarta (Indonesia) and its geodetic monitoring system. *Nat. Hazards* **2001**, *23*, 365–387. [CrossRef]
8. Notti, D.; Mateos, R.M.; Monserrat, O.; Devanthéry, N.; Peinado, T.; Roldán, F.J.; Fernández-Chacón, F.; Galve, J.P.; Lamas, F.; Azañón, J.M. Lithological control of land subsidence induced by groundwater withdrawal in new urban AREAS (Granada Basin, SE Spain). Multiband DInSAR monitoring. *Hydrol. Process.* **2016**, *30*, 2317–2331. [CrossRef]
9. Yastika, P.E.; Shimizu, N.; Abidin, H.Z. Monitoring of long-term land subsidence from 2003 to 2017 in coastal area of Semarang, Indonesia by SBAS DInSAR analyses using Envisat-ASAR, ALOS-PALSAR, and Sentinel-1A SAR data. *Adv. Space Res.* **2019**, *63*, 1719–1736. [CrossRef]
10. Fiaschi, S.; Tessitore, S.; Bonì, R.; Di Martire, D.; Achilli, V.; Borgstrom, S.; Ibrahim, A.; Floris, M.; Meisina, C.; Ramondini, M.; et al. From ERS-1/2 to Sentinel-1: Two decades of subsidence monitored through A-DInSAR techniques in the Ravenna area (Italy). *GIScience Remote Sens.* **2017**, *54*, 305–328. [CrossRef]
11. Pradhan, B.; Abokharima, M.H.; Jebur, M.N.; Tehrany, M.S. Land subsidence susceptibility mapping at Kinta Valley (Malaysia) using the evidential belief function model in GIS. *Nat. Hazards* **2014**, *73*, 1019–1042. [CrossRef]
12. Bianchini, S.; Solari, L.; Del Soldato, M.; Raspini, F.; Montalti, R.; Ciampalini, A.; Casagli, N. Ground Subsidence Susceptibility (GSS) mapping in Grosseto plain (Tuscany, Italy) based on satellite InSAR data using frequency ratio and fuzzy logic. *Remote Sens.* **2019**, *11*, 2015. [CrossRef]
13. Arabameri, A.; Saha, S.; Roy, J.; Tiefenbacher, J.P.; Cerda, A.; Biggs, T.; Pradhan, B.; Thi Ngo, P.T.; Collins, A.L. A novel ensemble computational intelligence approach for the spatial prediction of land subsidence susceptibility. *Sci. Total Environ.* **2020**, *726*, 138595. [CrossRef]
14. Abdollahi, S.; Pourghasemi, H.R.; Ghanbarian, G.A.; Safaeian, R. Prioritization of effective factors in the occurrence of land subsidence and its susceptibility mapping using an SVM model and their different kernel functions. *Bull. Eng. Geol. Environ.* **2019**, *78*, 4017–4034. [CrossRef]
15. Oh, H.J.; Lee, S. Integration of ground subsidence hazard maps of abandoned coal mines in Samcheok, Korea. *Int. J. Coal Geol.* **2011**, *86*, 58–72. [CrossRef]
16. Oh, H.J.; Lee, S. Assessment of ground subsidence using GIS and the weights-of-evidence model. *Eng. Geol.* **2010**, *115*, 36–48. [CrossRef]
17. Pradhan, B.; Oh, H.-J.; Buchroithner, M. Weights-of-evidence model applied to landslide susceptibility mapping in a tropical hilly area. *Geomat. Nat. Hazards Risk* **2010**, *1*, 199–223. [CrossRef]
18. Kim, K.D.; Lee, S.; Oh, H.J.; Choi, J.K.; Won, J.S. Assessment of ground subsidence hazard near an abandoned underground coal mine using GIS. *Environ. Geol.* **2006**, *50*, 1183–1191. [CrossRef]
19. Hu, B.; Zhou, J.; Wang, J.; Chen, Z.; Wang, D.; Xu, S. Risk assessment of land subsidence at Tianjin coastal area in China. *Environ. Earth Sci.* **2009**, *59*, 269–276. [CrossRef]
20. Zhi-xiang, T.; Pei-xian, L.; Li-li, Y.; Ka-zhong, D. Study of the method to calculate subsidence coefficient based on SVM. *Procedia Earth Planet. Sci.* **2009**, *1*, 970–976. [CrossRef]

21. Rahmati, O.; Falah, F.; Naghibi, S.A.; Biggs, T.; Soltani, M.; Deo, R.C.; Cerdà, A.; Mohammadi, F.; Tien Bui, D. Land subsidence modelling using tree-based machine learning algorithms. *Sci. Total Environ.* **2019**, *672*, 239–252. [CrossRef]

22. Tien Bui, D.; Shahabi, H.; Shirzadi, A.; Chapi, K.; Pradhan, B.; Chen, W.; Khosravi, K.; Panahi, M.; Bin Ahmad, B.; Saro, L. Land Subsidence Susceptibility Mapping in South Korea Using Machine Learning Algorithms. *Sensors* **2018**, *18*, 2464. [CrossRef] [PubMed]

23. Choi, J.K.; Kim, K.D.; Lee, S.; Won, J.S. Application of a fuzzy operator to susceptibility estimations of coal mine subsidence in Taebaek City, Korea. *Environ. Earth Sci.* **2010**, *59*, 1009–1022. [CrossRef]

24. Jaafari, A.; Panahi, M.; Pham, B.T.; Shahabi, H.; Bui, D.T.; Rezaie, F.; Lee, S. Meta optimization of an adaptive neuro-fuzzy inference system with grey wolf optimizer and biogeography-based optimization algorithms for spatial prediction of landslide susceptibility. *Catena* **2019**, *175*, 430–445. [CrossRef]

25. Park, I.; Choi, J.; Jin Lee, M.; Lee, S. Application of an adaptive neuro-fuzzy inference system to ground subsidence hazard mapping. *Comput. Geosci.* **2012**, *48*, 228–238. [CrossRef]

26. Lee, S.; Park, I.; Choi, J.K. Spatial prediction of ground subsidence susceptibility using an artificial neural network. *Environ. Manag.* **2012**, *49*, 347–358. [CrossRef]

27. Ambrožič, T.; Turk, G. Prediction of subsidence due to underground mining by artificial neural networks. *Comput. Geosci.* **2003**, *29*, 627–637. [CrossRef]

28. Oh, H.-J.; Syifa, M.; Lee, C.-W.; Lee, S. Land Subsidence Susceptibility Mapping Using Bayesian, Functional, and Meta-Ensemble Machine Learning Models. *Appl. Sci.* **2019**, *9*, 1248. [CrossRef]

29. Ferretti, A.; Prati, C.; Rocca, F. Permanent scatterers in SAR interferometry. *IEEE Trans. Geosci. Remote Sens.* **2001**, *39*, 8–20. [CrossRef]

30. Ferretti, A.; Savio, G.; Barzaghi, R.; Borghi, A.; Musazzi, S.; Novali, F.; Prati, C.; Rocca, F. Submillimeter accuracy of InSAR time series: Experimental validation. *IEEE Trans. Geosci. Remote Sens.* **2007**, *45*, 1142–1153. [CrossRef]

31. Hooper, A.; Bekaert, D.; Spaans, K.; Arikan, M. Recent advances in SAR interferometry time series analysis for measuring crustal deformation. *Tectonophysics* **2012**, *514–517*, 1–13. [CrossRef]

32. Hooper, A. A multi-temporal InSAR method incorporating both persistent scatterer and small baseline approaches. *Geophys. Res. Lett.* **2008**, *35*. [CrossRef]

33. Osmanoğlu, B.; Sunar, F.; Wdowinski, S.; Cabral-Cano, E. Time series analysis of InSAR data: Methods and trends. *ISPRS J. Photogramm. Remote Sens.* **2016**, *115*, 90–102. [CrossRef]

34. Osmanoğlu, B.; Dixon, T.H.; Wdowinski, S.; Cabral-Cano, E.; Jiang, Y. Mexico City subsidence observed with persistent scatterer InSAR. *Int. J. Appl. Earth Obs. Geoinf.* **2011**, *13*, 1–12. [CrossRef]

35. Haji-Aghajany, S.; Amerian, Y. Atmospheric phase screen estimation for land subsidence evaluation by InSAR time series analysis in Kurdistan, Iran. *J. Atmos. Solar Terr. Phys.* **2020**, *205*, 105314. [CrossRef]

36. Yang, C.; Zhang, F.; Liu, R.; Hou, J.; Zhang, Q.; Zhao, C. Ground deformation and fissure activity of the Yuncheng Basin (China) revealed by multiband time series InSAR. *Adv. Space Res.* **2020**, *66*. [CrossRef]

37. Tosi, L.; Da Lio, C.; Teatini, P.; Strozzi, T. Land subsidence in coastal environments: Knowledge advance in the Venice coastland by TerraSAR-X PSI. *Remote Sens.* **2018**, *10*, 1191. [CrossRef]

38. Kim, S.W.; Dixon, T.; Amelung, F.; Wdowinski, S. A time-series deformation analysis from TerraSAR-X SAR data over New Orleans, USA. In Proceedings of the 2011 3rd International Asia-Pacific Conference on Synthetic Aperture Radar, APSAR 2011, Seoul, Korea, 26–30 September 2011; pp. 832–833.

39. Zhao, Q.; Ma, G.; Wang, Q.; Yang, T.; Liu, M.; Gao, W.; Falabella, F.; Mastro, P.; Pepe, A. Generation of long-term InSAR ground displacement time-series through a novel multi-sensor data merging technique: The case study of the Shanghai coastal area. *ISPRS J. Photogramm. Remote Sens.* **2019**, *154*, 10–27. [CrossRef]

40. Imakiire, T.; Koarai, M. Wide-area land subsidence caused by "the 2011 off the Pacific Coast of Tohoku Earthquake". *Soils Found.* **2012**, *52*, 842–855. [CrossRef]

41. Hong, S.-J.; Baek, W.-K.; Jung, H.-S. Mapping Precise Two-dimensional Surface Deformation on Kilauea Volcano, Hawaii using ALOS2 PALSAR2 Spotlight SAR Interferometry. *Korean J. Remote Sens.* **2019**, *35*, 1235–1249. [CrossRef]

42. Jo, M.; Osmanoglu, B.; Jung, H.-S. Detecting Surface Changes Triggered by Recent Volcanic Activities at Kīlauea, Hawai'i, by using the SAR Interferometric Technique: Preliminary Report. *Korean J. Remote Sens.* **2018**, *34*, 1545–1553. [CrossRef]

43. Liu, P.; Li, Z.; Hoey, T.; Kincal, C.; Zhang, J.; Zeng, Q.; Muller, J.P. Using advanced inSAR time series techniques to monitor landslide movements in Badong of the Three Gorges region, China. *Int. J. Appl. Earth Obs. Geoinf.* **2012**, *21*, 253–264. [CrossRef]

44. Samsonov, S.; d'Oreye, N.; Smets, B. Ground deformation associated with post-mining activity at the French-German border revealed by novel InSAR time series method. *Int. J. Appl. Earth Obs. Geoinf.* **2013**, *23*, 142–154. [CrossRef]

45. Jung, H.C.; Kim, S.W.; Jung, H.S.; Min, K.D.; Won, J.S. Satellite observation of coal mining subsidence by persistent scatterer analysis. *Eng. Geol.* **2007**, *92*, 1–13. [CrossRef]

46. Ng, A.H.M.; Ge, L.; Li, X.; Abidin, H.Z.; Andreas, H.; Zhang, K. Mapping land subsidence in Jakarta, Indonesia using persistent scatterer interferometry (PSI) technique with ALOS PALSAR. *Int. J. Appl. Earth Obs. Geoinf.* **2012**, *18*, 232–242. [CrossRef]

47. BPS-Statistics of DKI Jakarta Province. *DKI Jakarta Province in Figures*; Division of Integration Processing and Statistics Dissemination BPS-Statistics of DKI Jakarta Province, Ed.; BPS-Statistics of DKI Jakarta Province: Jakarta, Indonesia, 2020; ISBN 978-602-0922-38-6.

48. Hudalah, D.; Firman, T. Beyond property: Industrial estates and post-suburban transformation in Jakarta Metropolitan Region. *Cities* **2012**, *29*, 40–48. [CrossRef]

49. Abidin, H.Z.; Andreas, H.; Gamal, M.; Gumilar, I.; Napitupulu, M.; Fukuda, Y.; Deguchi, T.; Maruyama, Y. Edi Riawan Land subsidence characteristics of the Jakarta basin (Indonesia) and its relation with groundwater extraction and sea level rise. *Groundw. Response Chang. Clim.* **2010**, 113–130. [CrossRef]

50. Arabameri, A.; Lee, S.; Tiefenbacher, J.P.; Ngo, P.T.T. Novel Ensemble of MCDM-Artificial Intelligence Techniques for Groundwater-Potential Mapping in Arid and Semi-Arid Regions (Iran). *Remote Sens.* **2020**, *12*, 490. [CrossRef]

51. Farr, T.G.; Rosen, P.A.; Caro, E.; Crippen, R.; Duren, R.; Hensley, S.; Kobrick, M.; Paller, M.; Rodriguez, E.; Roth, L.; et al. The Shuttle Radar Topography Mission. *Rev. Geophys.* **2007**, *45*, RG2004. [CrossRef]

52. Pourghasemi, H.R.; Beheshtirad, M. Assessment of a data-driven evidential belief function model and GIS for groundwater potential mapping in the Koohrang Watershed, Iran. *Geocarto Int.* **2015**, *30*, 662–685. [CrossRef]

53. Lee, S.; Park, I. Application of decision tree model for the ground subsidence hazard mapping near abandoned underground coal mines. *J. Environ. Manag.* **2013**, *127*, 166–176. [CrossRef]

54. Hooper, A. Persistent scatter radar interferometry for crustal deformation studies and modeling of volcanic deformation. Ph.D. Thesis, Stanford University, Stanford, CA, USA, 2006.

55. Wegnüller, U.; Werner, C.; Strozzi, T.; Wiesmann, A.; Frey, O.; Santoro, M. Sentinel-1 Support in the GAMMA Software. *Procedia Comput. Sci.* **2016**, *100*, 1305–1312. [CrossRef]

56. Werner, C.; Wegmüller, U.; Strozzi, T.; Wiesmann, A. GAMMA SAR and interferometric processing software. In Proceedings of the ERS—ENVISAT Symposium, Gothenburg, Sweden, 16–20 October 2000.

57. Crosetto, M.; Monserrat, O.; Cuevas-González, M.; Devanthéry, N.; Crippa, B. Persistent Scatterer Interferometry: A review. *ISPRS J. Photogramm. Remote Sens.* **2016**, *115*, 78–89. [CrossRef]

58. Hooper, A.; Segall, P.; Zebker, H. Persistent scatterer interferometric synthetic aperture radar for crustal deformation analysis, with application to Volcán Alcedo, Galápagos. *J. Geophys. Res. Solid Earth* **2007**, *112*, 1–21. [CrossRef]

59. Sousa, J.J.; Hooper, A.J.; Hanssen, R.F.; Bastos, L.C.; Ruiz, A.M. Persistent Scatterer InSAR: A comparison of methodologies based on a model of temporal deformation vs. spatial correlation selection criteria. *Remote Sens. Environ.* **2011**, *115*, 2652–2663. [CrossRef]

60. Pepe, A.; Bonano, M.; Zhao, Q.; Yang, T.; Wang, H. The Use of C-/X-Band Time-Gapped SAR Data and Geotechnical Models for the Study of Shanghai's Ocean-Reclaimed Lands through the SBAS-DInSAR Technique. *Remote Sens.* **2016**, *8*, 911. [CrossRef]

61. Floris, M.; Fontana, A.; Tessari, G.; Mulè, M. Subsidence Zonation Through Satellite Interferometry in Coastal Plain Environments of NE Italy: A Possible Tool for Geological and Geomorphological Mapping in Urban Areas. *Remote Sens.* **2019**, *11*, 165. [CrossRef]

62. Ren, H.; Feng, X. Calculating vertical deformation using a single InSAR pair based on singular value decomposition in mining areas. *Int. J. Appl. Earth Obs. Geoinf.* **2020**, *92*, 102115. [CrossRef]

63. Freund, Y.; Schapire, R.E. A Decision-Theoretic Generalization of On-Line Learning and an Application to Boosting. *J. Comput. Syst. Sci.* **1997**, *55*, 119–139. [CrossRef]

64. Kadavi, P.; Lee, C.-W.; Lee, S. Application of Ensemble-Based Machine Learning Models to Landslide Susceptibility Mapping. *Remote Sens.* **2018**, *10*, 1252. [CrossRef]
65. Sprenger, M.; Schemm, S.; Oechslin, R.; Jenkner, J. Nowcasting foehn wind events using the AdaBoost machine learning algorithm. *Weather Forecast.* **2017**, *32*, 1079–1099. [CrossRef]
66. Friedman, J.; Hastie, T.; Tibshirani, R. Additive logistic regression: A statistical view of boosting. *Ann. Stat.* **2000**, *28*, 337–407. [CrossRef]
67. Zhang, G.; Fang, B. LogitBoost classifier for discriminating thermophilic and mesophilic proteins. *J. Biotechnol.* **2007**, *127*, 417–424. [CrossRef]
68. Lee, S. Application of logistic regression model and its validation for landslide susceptibility mapping using GIS and remote sensing data. *Int. J. Remote Sens.* **2005**, *26*, 1477–1491. [CrossRef]
69. Erener, A.; Mutlu, A.; Sebnem Düzgün, H. A comparative study for landslide susceptibility mapping using GIS-based multi-criteria decision analysis (MCDA), logistic regression (LR) and association rule mining (ARM). *Eng. Geol.* **2016**, *203*, 45–55. [CrossRef]
70. Ozdemir, A.; Altural, T. A comparative study of frequency ratio, weights of evidence and logistic regression methods for landslide susceptibility mapping: Sultan mountains, SW Turkey. *J. Asian Earth Sci.* **2013**, *64*, 180–197. [CrossRef]
71. David, W.; Hosmer, J.; Lemeshow, S.; Sturdivant, R.X. *Applied Logistic Regression*, 3rd ed.; John Wiley & Sons, Ltd.: Hoboken, NJ, USA, 2013.
72. Pham, B.T.; Tien Bui, D.; Prakash, I.; Dholakia, M.B. Hybrid integration of Multilayer Perceptron Neural Networks and machine learning ensembles for landslide susceptibility assessment at Himalayan area (India) using GIS. *Catena* **2017**, *149*, 52–63. [CrossRef]
73. Gardner, M.W.; Dorling, S.R. Artificial neural networks (the multilayer perceptron)—A review of applications in the atmospheric sciences. *Atmos. Environ.* **1998**, *32*, 2627–2636. [CrossRef]
74. Kim, S.-W.; Wdowinski, S.; Dixon, T.H.; Amelung, F.; Won, J.-S.; Kim, J.W. InSAR-based mapping of surface subsidence in Mokpo City, Korea, using JERS-1 and ENVISAT SAR data. *Earth Planets Space* **2008**, *60*, 453–461. [CrossRef]
75. Fawcett, T. An introduction to ROC analysis. *Pattern Recognit. Lett.* **2006**, *27*, 861–874. [CrossRef]
76. Van Leijen, F.J. Persistent Scatterer Interferometry based on geodetic estimation theory. Ph.D. Thesis, Technische Universiteit Delft, Delft, The Netherlands, 2014.
77. Colesanti, C.; Ferretti, A.; Novali, F.; Prati, C.; Rocca, F. SAR monitoring of progressive and seasonal ground deformation using the permanent scatterers technique. *IEEE Trans. Geosci. Remote Sens.* **2003**, *41*, 1685–1701. [CrossRef]
78. Smith, R.G.; Knight, R.; Chen, J.; Reeves, J.A.; Zebker, H.A.; Farr, T.; Liu, Z. Estimating the permanent loss of groundwater storage in the southern San Joaquin Valley, California. *Water Resour. Res.* **2017**, *53*, 2133–2148. [CrossRef]
79. Reeves, J.A.; Knight, R.; Zebker, H.A.; Kitanidis, P.K.; Schreüder, W.A. Estimating temporal changes in hydraulic head using InSAR data in the San Luis Valley, Colorado. *Water Resour. Res.* **2014**, *50*, 4459–4473. [CrossRef]
80. Erban, L.E.; Gorelick, S.M.; Zebker, H.A. Groundwater extraction, land subsidence, and sea-level rise in the Mekong Delta, Vietnam. *Environ. Res. Lett.* **2014**, *9*. [CrossRef]
81. Erban, L.E.; Gorelick, S.M.; Zebker, H.A.; Fendorf, S. Release of arsenic to deep groundwater in the Mekong Delta, Vietnam, linked to pumping-induced land subsidence. *Proc. Natl. Acad. Sci. USA* **2013**, *110*, 13751–13756. [CrossRef] [PubMed]

Publisher's Note: MDPI stays neutral with regard to jurisdictional claims in published maps and institutional affiliations.

 *remote sensing*

Article

Integration of InSAR Time-Series Data and GIS to Assess Land Subsidence along Subway Lines in the Seoul Metropolitan Area, South Korea

Muhammad Fulki Fadhillah, Arief Rizqiyanto Achmad and Chang-Wook Lee *

Division of Science Education, Kangwon National University, Gangwon-do, Chuncheon-si 24341, Korea;
fulkifadhillah@kangwon.ac.kr (M.F.F.); ariefrizqiyanto@kangwon.ac.kr (A.R.A.)
* Correspondence: cwlee@kangwon.ac.kr

Received: 9 September 2020; Accepted: 22 October 2020; Published: 25 October 2020

Abstract: The aims of this research were to map and analyze the risk of land subsidence in the Seoul Metropolitan Area, South Korea using satellite interferometric synthetic aperture radar (InSAR) time-series data, and three ensemble machine-learning models, Bagging, LogitBoost, and Multiclass Classifier. Of the types of infrastructure present in the Seoul Metropolitan Area, subway lines may be vulnerable to land subsidence. In this study, we analyzed Persistent Scatterer InSAR time-series data using the Stanford Method for Persistent Scatterers (StaMPS) algorithm to generate a deformation time-series map. Subsidence occurred at four locations, with a deformation rate that ranged from 6–12 mm/year. Subsidence inventory maps were prepared using deformation time-series data from Sentinel-1. Additionally, 10 potential subsidence-related factors were selected and subjected to Geographic Information System analysis. The relationship between each factor and subsidence occurrence was analyzed by using the frequency ratio. Land subsidence susceptibility maps were generated using Bagging, Multiclass Classifier, and LogitBoost models, and map validation was carried out using the area under the curve (AUC) method. Of the three models, Bagging produced the largest AUC (0.883), with LogitBoost and Multiclass Classifier producing AUCs of 0.871 and 0.856, respectively.

Keywords: Seoul; synthetic aperture radar; land subsidence; GIS; machine learning; time-series

---

1. Introduction

Land subsidence is a threat faced by big cities with extensive development that can negatively impact the environment, social systems, and the economy [1]. Subsidence occurs due to geological causes or anthropogenic processes such as massive urban development, infrastructure development [2,3], tunneling [4–6], water extraction [7–9], and earthquakes [10]. Subsidence has been observed in several metropolitan cities, including Mexico City [11], Shanghai [12], and Jakarta [13–15]. The Seoul Metropolitan Area is the center of governance, commerce, and culture in South Korea. It has been extensively developed and is the most densely populated city in Asia [16]. Industrial development and economic growth have led to city developments such as the expansion of subway lines and the construction of many structures and buildings [17]. By examining the potential effects of land subsidence, monitoring of land deformation could be the first step of a mitigation process. Seoul, which has a high population density and extensive developments, is extremely vulnerable to land subsidence. Given the severe negative impacts of subsidence, it necessary to elucidate the factors that cause land subsidence from integrated observations, to assess damage risks and prevent damage to roads, bridge, railways, and other infrastructure.

Monitoring land deformation is essential for reducing losses due to subsidence and developing a sound mitigation plan. With advancements in knowledge and technology, monitoring techniques have improved greatly. Previously, measurements related to land subsidence could be taken using traditional geodetic [15] and leveling methods, which provide precise measurements, but these methods are inefficient and costly compared with satellite-based methods, which cover broader areas and are more cost-efficient. These days, more studies are using satellite-based synthetic aperture radar (SAR) to monitor land deformation. SAR is an advanced remote-sensing technology that has been used worldwide in several applications; it is considered active remote sensing as the sensor emits its own microwaves before recording the backscattered waves [18]. In addition, another advantage of SAR is its flexibility in acquiring data, as it can operate under any weather conditions and all day [19]. Thus, because of its operational flexibility, SAR is considered reliable for damage assessment and risk analysis.

Recently, there has been growing interest in monitoring ground deformation using differential interferometric SAR (DInSAR), which is used to estimate such small deformation. However, this method is often restrained by temporal decorrelation and atmospheric disturbance. Those problems are difficult to eliminate when estimating the representative interferograms of ground deformations [20]. To solve this problem, a long-term InSAR processing approach using a series of data, generally recognized as time-series InSAR analysis. Time-series analysis has been widely used for ground-monitoring projects, such as monitoring urban areas, groundwater, and land subsidence [21–23]. An example of InSAR time-series methods that have been developed in recent years is Persistent Scatterer InSAR (PS-InSAR). With time-series data accumulated over long periods of monitoring, deformation can be measured over broader areas, and the occurrence of deformation in an area can be studied with more precision compared to data from conventional InSAR methods. In addition, a prominent advantage of time-series analysis is a reduction in issues related to InSAR, such as temporal decorrelation and atmospheric interference.

In PS-InSAR, multiple large SAR images of the same area are processed, which can extract a number of persistent scatterers (PS) [24]. This method focuses on measuring the level of deformation associated with each of the persistent scatterers, which is a point of high density within an interferogram from a single main image. Therefore, this technique was developed for urban areas, which have more stable scatterers than mountains or forests with distributed scatterers [25]. In recent years, many PSI approaches have been developed and applied in many cases, such as PS-InSAR [26–28], Stable Point Network (SPN) [29], Interferometric Point Target Analysis (IPTA) [30–32], and Stanford Method of Persistent Scatterers (StaMPS) [33,34]. This method has been applied to measure deformation in several locations such as West Macedonia, Greece [35]; Nile Delta, Egypt [23]; Tuscany and Northern Apennines, Italy [20,36,37]; Mexico City, Mexico [11]; and Guangzhou, China [30,38].

In general, susceptibility maps are generated as part of land-subsidence mapping and modeling. These maps are used to predict and examine areas that are highly vulnerable to land subsidence and are important as part of the initial information used to prevent future land-subsidence events, assuming the same conditions will trigger land subsidence in the future [39]. Land-subsidence analysis largely focuses on elucidating the sources of subsidence, methods of evaluation, understanding subsidence events on a conceptual level, and mapping [40]. Remote sensing and Geographical Information System (GIS) analysis are commonly employed in hazard studies because of the efficiency of these techniques in data collection and analysis [41]. The utility of several GIS methods in terms of generating subsidence-susceptibility maps has been compared using statistical tools such as frequency ratio and weight of evidence [6,37]. Artificial neural networks, random forests, and fuzzy logic have also been applied to assess methods of predicting subsidence susceptibility [42–44]. Furthermore, machine-learning techniques have recently attracted attention in the environmental-modeling research community as they are highly efficient and generate improved outcomes.

Previous studies related to land subsidence in Seoul related to the risk analysis of ground subsidence around railways using ANN modeling [45]. However, studies related to land subsidence in Seoul are needed to map the potential for land subsidence and understand the causes that can lead to land subsidence. In this study, we aimed to examine ground deformation within the 2017–2020 time period in the Seoul Metropolitan Area using C-band data from the Sentinel-1 satellite. The velocity of land subsidence and the associated land deformation was investigated via Stanford Method for Persistent Scatterers (StaMPS), and a deformation time-series map was generated. The map was used as input data for a land-subsidence inventory map, which was then used to generate a land-subsidence-susceptibility map. GIS analysis was employed to complement observations of land subsidence in the urban area. Thereafter, the performance of the meta-ensemble methods in terms of land-subsidence-susceptibility modeling was assessed. The LogitBoost, Multiclass Classifier, and Bagging functional models were used as they have been shown to improve the performance of predictive models in several cases of susceptibility map [43,46,47]. Bagging more effectively reduces the bias compared to other ensembles, while LogitBoost is used to solve the overfitting problem [43,48]. Model performance in terms of predicting land-subsidence susceptibility was assessed using training and testing datasets, receiver operating characteristic (ROC), and the area under the curve (AUC). The area under the curve of the roc represents the validation of systems and its ability to predict the correct occurrence or non-occurrence of land subsidence events [49]. The ROC graph is a technique for visualizing, organizing, and selecting classifier based on their performance. Then, the area under the curve is a common method to calculate the area under the curve to compare the classifiers and convert it to the scalar values which represent the performance [50]. By linking time-series data with GIS analysis, the research findings will increase our understanding of crucial factors affecting ground surface areas. The map of ground-deformation risk in an urban area generated in this study reflects the association between subsidence risk and risk factors; this information would help identify areas of high risk and develop environmental action plans and policies.

2. Materials & Methods

2.1. Study Area

Seoul is the capital city of South Korea. It is located in the midwestern region of the Korean Peninsula at 126°59′40″E and 37°33′59″N and covers an area of 605.5 km^2 [51]. The Han River, which is one of the largest rivers crossing Seoul, divides the city into north and south areas. Seoul has a population of approximately 10 million people with a density of 16,364 people/km^2, making it one of the most populous metropolitan cities in Asia [52]. The geological setting of Seoul consists of Jurassic granite, Precambrian metamorphic rocks (gneiss and schist), and Quaternary alluvium. Predominantly, coarse-grained, sandy alluvium sequence (<20 m thick) occurs along the Han River and its tributaries [53]. The alluvium is mainly distributed along the Han River and its tributaries, it is composed of coarse- to fine-grained sediments, often with high permeability. The alluvium and soil tend to be thicker close to the river, particularly its lower reaches, and thinner in mountain area [54]. In this study area, there are two types of aquifer unconsolidated alluvium aquifers and bedrock aquifers [17]. The alluvial aquifers are dominantly composed of silt and fine to coarse sands are appearance along the Han river and tributaries. The bedrock aquifers are mainly composed of fractured gneiss, schist, and granite.

As a metropolitan city that has experienced urbanization in recent years, Seoul has experienced many developments such as office, business, and residential buildings. This has an impact on increasing the density of the building in this area. The use of groundwater and other utilities in densely populated areas will have an impact on the weakening of soil conditions in these areas. This condition can indirectly lead to subsidence which can cause many losses, especially in areas with high population density such as Seoul.

Together with the increase in population, the economy has grown quickly, followed by industrial development. To meet the needs of the city inhabitants, Seoul undertook massive developments, including infrastructure, buildings, and transportation networks. At the end of 2019, a total of 23 rapid transit, light metro, commuter rail, and airport rail lines had been integrated into the Seoul Metropolitan Subway system [55]. This system operates in the Seoul Metropolitan Area, including Incheon and some satellite cities in Gyeonggi Province. Several regional lines such as those in Chungnam and Gangwon provinces are also connected to this system. Figure 1 has shown a map of the subway line that has been operating and in the process of construction in the Seoul Metropolitan Area. New transportation routes have since been added, such as the Gimpo Gold-Line in 2019, and the Line 7 and Line 5 extensions to Hanam City is currently under construction and slated to open in 2020. To improve connectivity in this metropolitan area, several future subway lines (until 2028) are still being planned.

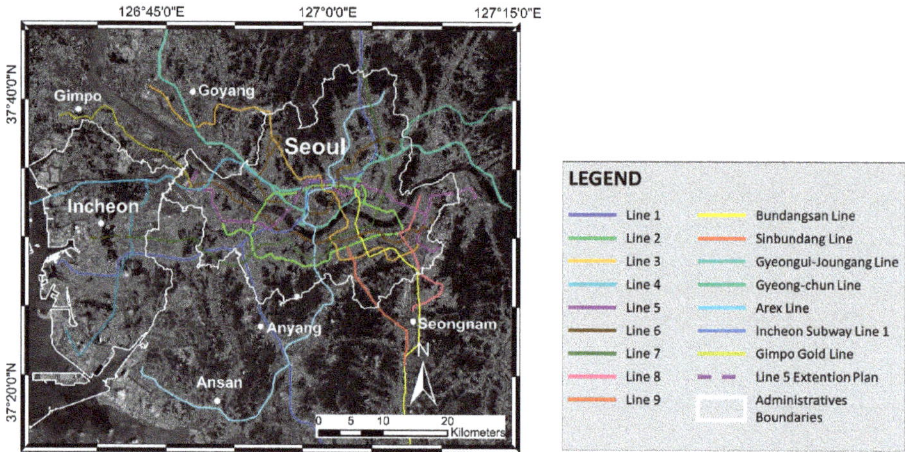

Figure 1. Optical image of Seoul (gray) captured by Sentinel-2 on 19 June 2019 with subway lines superimposed.

In densely populated areas like Seoul, ground subsidence can cause much higher casualties and property damages than in lesser populated areas. For this reason, it is of utmost importance to conduct complete monitoring on the cases of ground subsidence to prevent damages on roads, railroads, and other infrastructures.

2.2. SAR Datasets

In this study, SAR data from Sentinel-1B was used to generate representations of surface deformation. SAR images derived from C-band data can be used to map surface deformation over broad areas while providing time-consistent ground-deformation data. Sentinel-1B has an acquisition cycle of 12 days. We used 93 SAR scenes from descending tracks. The descending datasets are listed in Table 1, the reference date with zero delta day and zero perpendicular baselines from the descending track is shown on October 11, 2018 as the reference date are shown in bold text.

2.3. StaMPS Processing

One of the known methods for generating time-series data on surface deformation is StaMPS, as it can be used for analysis on the urban area like this study area. This method is commonly recognized on man-made objects such as buildings, infrastructure, and roads in urbanized areas like Seoul. Another major advantage of this method is that it does not require a prior deformation model, thus allowing analysis of different regions and several deformation causes [24].

Table 1. Acquisition dates of data from the Sentinel-1 satellite in descending tracks, the reference date shown in bold text.

No.	Acquisition Date (yyyymmdd)	Days	$B_\perp$(m)[1]	No.	Acquisition Date (yyyymmdd)	Days	$B_\perp$(m)[1]	No.	Acquisition Date (yyyymmdd)	Days	$B_\perp$(m)[1]
1	20170302	−588	78	32	20180414	−180	102	63	20190421	192	125
2	20170314	−576	128	33	20180426	−168	67	64	20190503	204	187
3	20170326	−564	128	34	20180508	−156	39	65	20190515	216	97
4	20170407	−552	91	35	20180520	−144	17	66	20190527	228	91
5	20170419	−540	61	36	20180601	−132	−2	67	20190620	252	82
6	20170501	−528	83	37	20180613	−120	117	68	20190702	264	160
7	20170513	−516	117	38	20180625	−108	87	69	20190714	276	117
8	20170525	−504	82	39	20180707	−96	65	70	20190807	300	41
9	20170606	−492	121	40	20180719	−84	93	71	20190819	312	100
10	20170618	−480	49	41	20180731	−72	24	72	20190831	324	87
11	20170630	−468	14	42	20180812	−60	67	73	20190912	336	99
12	20170712	−456	91	43	20180824	−48	69	74	20190924	348	112
13	20170805	−444	129	44	20180905	−36	77	75	20191006	360	55
14	20170817	−420	12	45	20180917	−24	62	76	20191018	372	103
15	20170910	−408	78	46	20180929	−12	74	77	20191030	384	64
16	20170922	−384	134	**47**	**20181011**	**0**	**0**	78	20191111	396	132
17	20171004	−372	105	48	20181023	12	163	79	20191123	408	126
18	20171016	−360	96	49	20181104	24	109	80	20191205	420	66
19	20171028	−348	96	50	20181116	36	106	81	20191217	432	110
20	20171109	−336	71	51	20181128	48	70	82	20191229	444	163
21	20171121	−324	126	52	20181210	60	122	83	20200110	456	176
22	20171203	−312	150	53	20181222	72	190	84	20200203	480	101
23	20171215	−300	148	54	20190103	84	119	85	20200215	492	61
24	20171227	−288	103	55	20190115	96	58	86	20200227	504	70
25	20180108	−276	102	56	20190127	108	96	87	20200310	516	158
26	20180201	−264	159	57	20190208	120	141	88	20200322	528	141
27	20180213	−240	166	58	20190220	132	174	89	20200403	540	80
28	20180225	−228	44	59	20190304	144	157	90	20200415	552	46
29	20180309	−216	−7	60	20190316	156	31	91	20200427	564	63
30	20180321	−204	−35	61	20190328	168	29	92	20200509	576	72
31	20180402	−192	126	62	20190409	180	68	93	20200521	588	56

[1] $B_\perp$: Perpendicular Baseline.

At the start of PSI-StaMPS analysis, the interferogram process was begun to generate a couple of interferogram images from the 93 SAR scenes in descending track. Prior to interferogram generation, SAR data underwent a co-registration process, in which two SAR images were aligned to subpixel accuracy for accurate determination and noise reduction to form interferometric pairs. The SAR images were then resampled such that the slave images matched the master image. When the co-registration process was complete, the co-registered images were cropped to focus on the study area before the interferogram generation process began. During the interferogram processing stage, a topographical phase was generated. Once the interferogram images were generated, the topographic phase was subtracted from the interferogram using Shuttle Radar Topography Mission digital elevation model (SRTM DEM) as the reference [56]. After the topographic phase was removed, a DInSAR phase was generated, which contained only the deformation phase.

For StaMPS processing, SAR data on 11 October 2018 was chosen as a master image and generated 92 interferograms from the descending track. After generating the interferograms, the StaMPS module was used to calculate the displacements of persistent scatterers. To begin the StaMPS process, phase stability estimation was used to select a subset of pixels based on amplitude analysis. Then, phase stability for each pixel was estimated through phase analysis [57]. Once the phase noise associated with all selected persistent scatterer (PS) pixels was estimated, the selected PS pixels were weeded out to separate the persistent points and noise, then the wrapped phase of the selected pixels was corrected for spatially-uncorrelated look-angle errors in the DEM. After correction, the corrected phase could now be unwrapped, and the PS output was generated. The parameter of the StaMPS process is shown in Table 2. Upon completion of the StaMPS process, PS results were plotted in a time-series map and a mean deformation from the line of sight (LOS) map [58,59]. The mean deformation map is converted into vertical deformation data by assuming the horizontal deformation is very small compared to the vertical deformation that causes by land subsidence [60,61].

In recent studies, the vertical deformation was used to monitor land subsidence in several places, therefore horizontal deformation in this study can be assumed as negligible and converted into vertical deformation value [13,14]. The result of vertical deformation will be assigned as a negative value from initial ground-level observation, which indicates the land subsidence measure vertically on that point. The vertical deformation can be calculated using this Equation (1) as follows [62]:

$$V = \frac{d_{LOS}}{\cos \theta} \tag{1}$$

where d_{LOS} is the deformation in line of sight and θ is the incident angle.

Table 2. The parameter in StaMPS processing.

Parameter	Value
DEM	SRTM 1 arc second
Maximum DEM error	20 m
Band-pass phase filter grid size	50
Band-pass phase filter low-pass cutoff	800
Band-pass phase filter low-pass α	1
Band-pass phase filter low-pass β	0.3
Unwrapping algorithm	3D unwrapping
Unwrapping grid cell size	100
Unwrapping Gaussian width	8σ

2.4. Generation of Susceptibility Map

The workflow to generate land subsidence susceptibility maps, using machine learning algorithms, is illustrated in Figure 2, the summary of the methodology is as follows:

1. The land subsidence inventory was generated by analyzing Sentinel-1 SAR datasets from 2017 to 2020 from descending tracks using the time-series InSAR technique based on StaMPS algorithms.
2. In order to generate land susceptibility maps, the training and test datasets were prepared by randomly divided the persistent scatterers (PS) points of time series into 50% of training data and 50% of testing datasets to validate the land subsidence susceptibility map. Training data is used to train the machine learning to predict subsidence in our land subsidence susceptibility model. Besides, test data is used to measure the performance, of the algorithm that we used to make the land subsidence susceptibility model. This preparation method of training and testing datasets was used in several studies of land subsidence susceptibility which has optimal results [6,63,64].
3. Preparation of land subsidence conditioning factors: Spatial correlation analysis was applied to assess each factor before the land-subsidence model was generated. In the spatial correlation analysis, the spatial relationship between historical subsidence events, and each factor was examined [65]. Spatial correlation analysis was also used to investigate the weight of each factor class to assess the strength of the relationship between each factor class and subsidence occurrence. Frequency ratios were calculated to reflect spatial correlations by calculating the proportion of cells in which subsidence occurred in each class; then, factors were reclassified. Frequency ratios have been commonly used to determine spatial correlations [40,42,66]. Here, each frequency ratio represents the quantitative relationship between subsidence in a selected class and all subsidence in the area for all classes as a percentage of the entire map [67]. If the ratio is greater than one, the relationship between subsidence and the factor class is considered strong. By contrast, if the ratio is less than one, the spatial relationship is weak [40].
4. Generating land subsidence susceptibility map: in this step, we constructed a land subsidence susceptibility map using Bagging, LogitBoost, and Multiclass Classifier algorithms. The land subsidence conditioning factors that consist of frequency ratio values.

5. After the land subsidence susceptibility map was generated, all susceptibility maps were evaluated using ROC analysis.

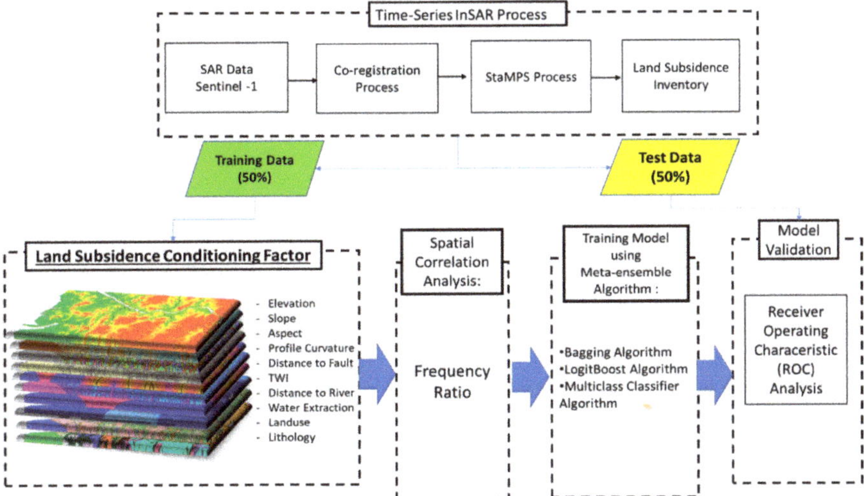

Figure 2. The workflow illustration.

2.4.1. Bagging

Bagging is a commonly used meta-algorithm that was developed to enhance the stability and accuracy of the machine-learning algorithms used in statistical classification and regression [43]. Bagging was one of the earliest ensemble techniques that used the bootstrap sampling method [68]. The bootstrap method entails apparent random sampling with replacements to generate more than one sample that shapes a training set. Each generated subset is used to assemble a decision tree, with all trees aggregated later into the final model. This improves class accuracy by reducing the variance of class error. We used Bagging to obtain a much improved and more accurate land subsidence model because this algorithm performs well in predicting land subsidence susceptibility, as it is sensitive to small adjustments in the training data and consequently [43,46]. Bagging ensembles more effectively reduce uncertainty and bias compared to other ensembles [69]. In addition, this algorithm is capable of reflecting complex non-linear interaction between land subsidence and related factors, although it lacks a statistical significance test which can limit quantitative hypothesis testing [43]. Bagging first uses a classifier to reduce variance, then carries out classification and regression by relying on bottom-up learning.

2.4.2. LogitBoost

LogitBoost is a boosting algorithm developed by Friedman et al., [70] to reduce bias and variance. The LogitBoost algorithm was modified from AdaBoost, which was the commonly boosting method for handling noisy data that execute additive logistic regression with least-square fits for individual class [48,71]. LogitBoost reduces training errors and enhances classification accuracy [72] by using additive logistic regression for classification with a base-learning regression scheme and an ability to perform multiclass classification. The land subsidence-inventory map was divided into two classes—subsidence occurrence and subsidence non-occurrence—using the following equation [71]:

$$Lc(c) = \sum_{i=1}^{D} \beta_i x_i + \beta_0 \tag{2}$$

where D is the number of landslide-dependent factors and β_i is the coefficient of the *i*-th component within input vector x. Probabilities were constructed using the linear logistic regression method, as follows:

$$p\left(\frac{C}{x}\right) = \exp(Lc(x)) / \sum_{C'=1}^{C} \exp(Lc'(x)) \tag{3}$$

where C is the number of classes and the least-square fit Lc(x) is resolved such $\sum_{C=1}^{C} L_C^C(x) = 0$ to set up the least number of instances per node of the logistic model trees.

2.4.3. Multiclass Classifier

Multiclass Classifier is a meta-classifier that is used to process multiclass datasets with two-class classifiers. It is efficient at applying error-correcting output codes to enhance accuracy [47]. In the field of machine learning, multiclass classification can classify events into one of three or more classes. Although several classification algorithms can work with more than two classes with the aid of natural binary algorithms, the conversion to multinomial classifiers requires the use of several strategies. Multiclass classification techniques can be divided into categories such as transformation to binary, extension from binary, and hierarchical classification [73].

2.5. Factors Related to Land Subsidence

The increase in land subsidence occurrence in megacities mainly due to lowered groundwater levels and the presence of heavy buildings [17]. Excessive use of groundwater has an impact on decreasing pore pressure on the soil, coupled with the presence of heavy buildings that lead to further soil compaction [74,75]. A large amount of groundwater leaked, and dewatering accompanies a decrease in groundwater levels. Those conditions weaken the surrounding land and lead to subsidence occurrence. In addition, based on a report by the Seoul government which has conducted field investigations, one of the causes of subsidence is excessive groundwater use and damage to water utilities and sewage [3,17].

A combination of several environmental factors can influence land-subsidence susceptibility. The training data and test data point was chosen from the land subsidence inventory map as shown in Figure 3a. Here, we investigated 10 subsidence-related factors (Figure 3b–k) and evaluated the correlation between each factor and land-subsidence occurrence as shown in Table 3 below. In the previous studies, land subsidence conditioning factors such as altitude, slope, aspect, plan curvature, profile curvature, lithology, distance to the river, land use, normalized differential vegetation index (NDVI), piezometric data (groundwater drawdown) have been used with the main cause of subsidence in Iran being groundwater drawdown [49]. Another study has evaluated several factors mentioned before to identify land subsidence in the mining area in Malaysia [40]. Reclassification was employed to place subsidence related-factors into several classes using the quantile method to objectively identify and analyze the effect of each class using a specific range of values. The quantile classification method can solve unbalanced distribution by focusing on the equality of domain grids [76]. Thus, the range of each class is automatically determined based on the quantile method.

The derivative feature from the digital elevation model (DEM) contains hydrogeological and topographic conditions such as hydrological zone response, concentration, and containment of runoff volume in the landscape, which directly or indirectly affect the occurrence of land subsidence. The topography features such as elevation, slope, aspect, topographic wetness index (TWI), and profile curvature (Figure 3b–f) data were extracted using SRTM DEM 1 arc second. This feature has been widely used as conditional factors in the land subsidence susceptibility model [37,43,49]. Elevation has a role as a bridge between lithology and rain characteristics in the area. A higher area has a lower probability of additional precipitation than a lower area which has the potential to have high precipitation [49]. The elevation refers to the height of the study area which varies between 0–813 m.

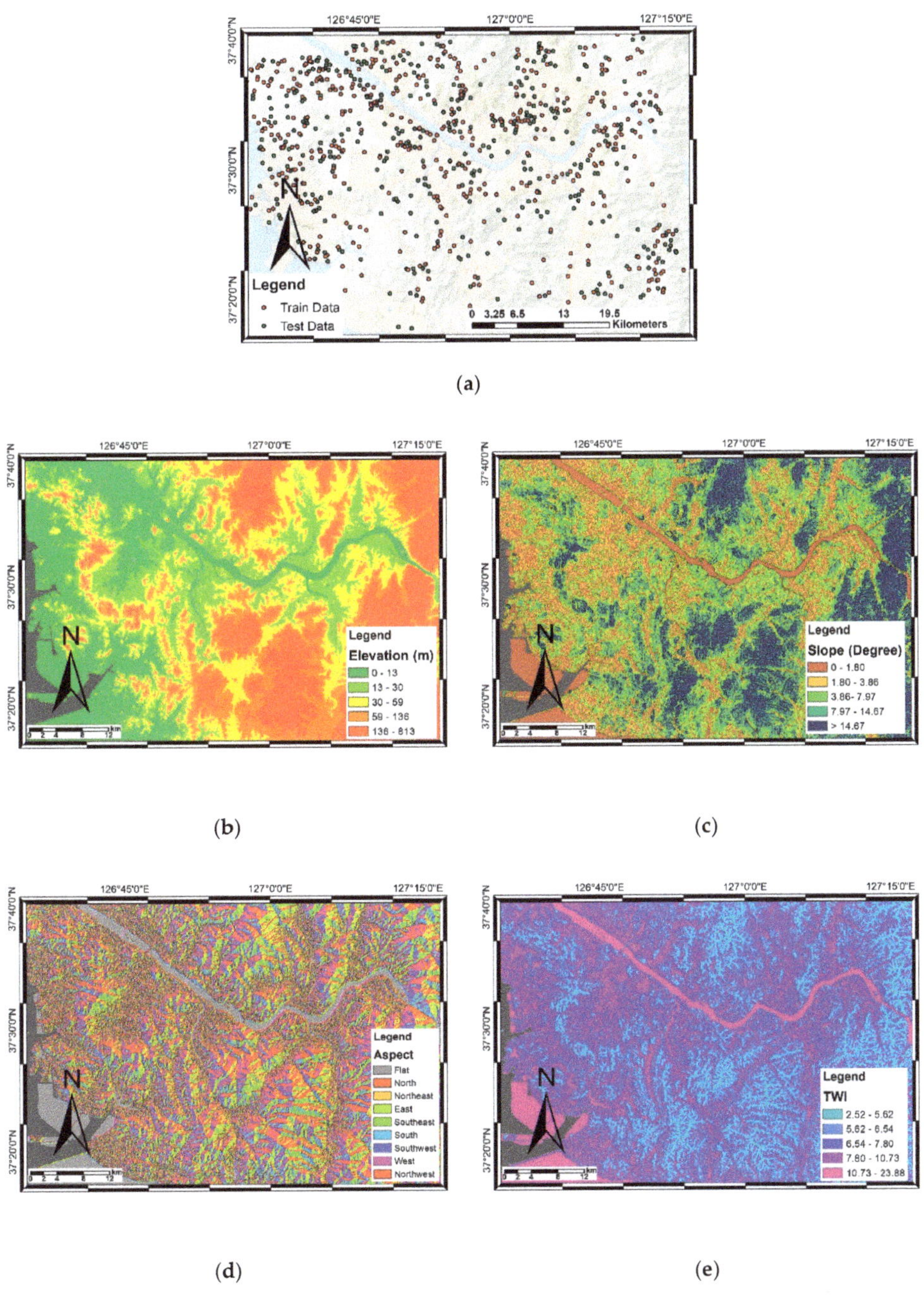

Figure 3. *Cont.*

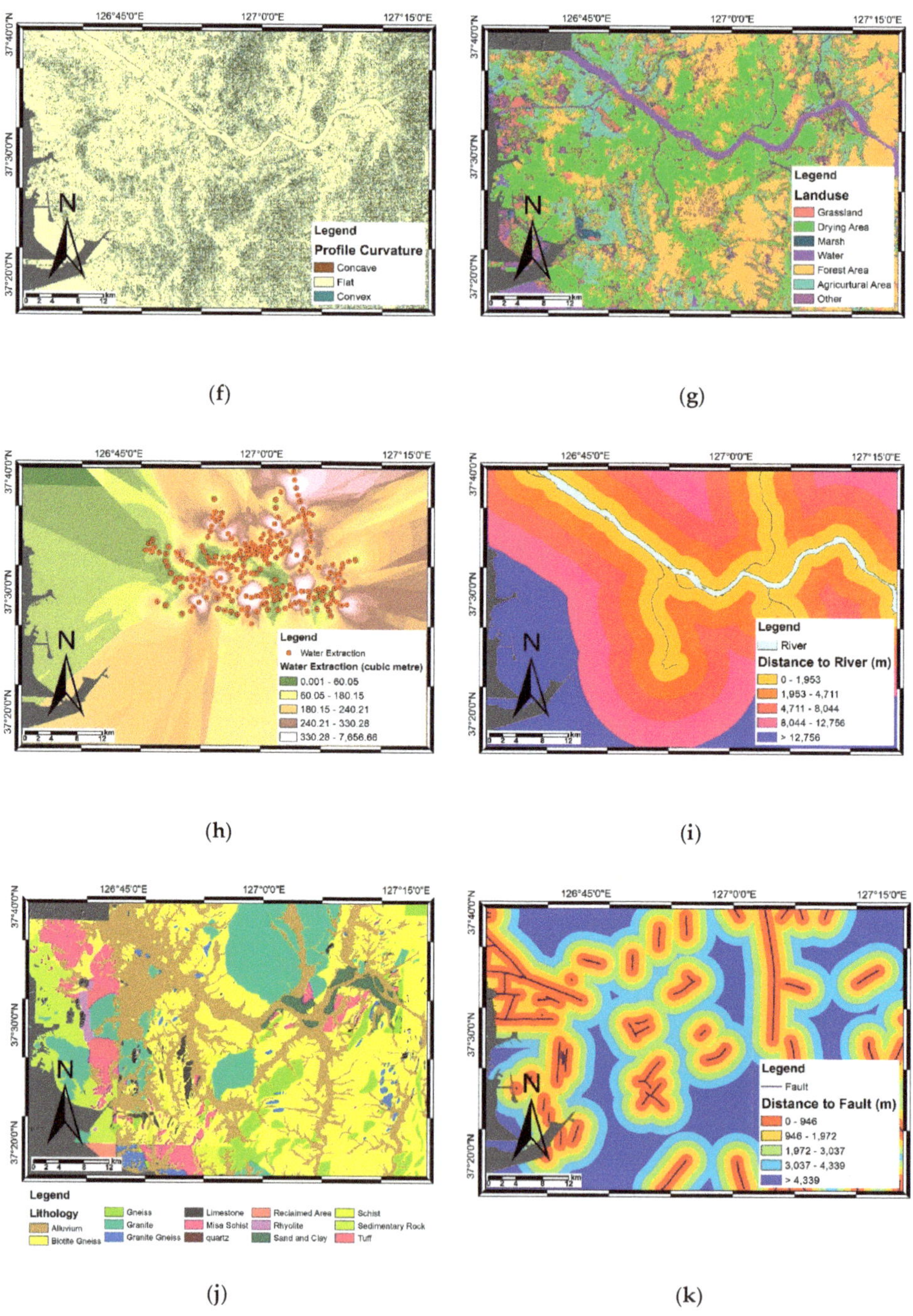

Figure 3. Train-test data point and spatial maps of 10 land-subsidence-related factor: train-test data point (**a**), elevation (**b**), slope (**c**), aspect (**d**), topographic wetness index (TWI) (**e**), profile curvature (**f**), land use (**g**), groundwater extraction (**h**), distance to river (**i**), lithology (**j**), and distance to fault (**k**).

Table 3. Frequency Ratio calculation.

Conditioning Factor	Class/Category	Ratio Each Class	Ratio of Occurrence	FR
Elevation	0–13	0.205	0.292	1.424
	13–30	0.206	0.355	1.722
	30–59	0.198	0.228	1.147
	59–136	0.197	0.100	0.510
	136–813	0.194	0.025	0.130
Aspect	Flat	0.066	0.071	1.071
	North	0.113	0.117	1.029
	Northeast	0.127	0.122	0.958
	East	0.125	0.132	1.059
	Southeast	0.128	0.135	1.053
	South	0.125	0.122	0.972
	Southwest	0.135	0.125	0.925
	West	0.124	0.121	0.976
	Northwest	0.056	0.056	0.999
Profile	concave	0.085	0.041	0.480
	flat	0.821	0.900	1.096
	convex	0.094	0.059	0.629
Slope	0–1.8	0.132	0.193	1.460
	1.8–3.86	0.218	0.350	1.601
	3.86–7.97	0.216	0.275	1.273
	7.97–14.67	0.216	0.142	0.656
	> 14.67	0.217	0.040	0.185
Topographic Wetness Index	2.52–5.62	0.207	0.072	0.347
	5.62–6.54	0.224	0.184	0.823
	6.54–7.80	0.228	0.264	1.157
	7.80–10.73	0.216	0.321	1.484
	10.73–23.88	0.125	0.159	1.270
Land use	Drying Area	0.318	0.658	2.069
	Agriculture Area	0.078	0.073	0.929
	Forest Area	0.316	0.049	0.157
	Grassland	0.125	0.148	1.183
	Marsh	0.023	0.007	0.308
	Other	0.058	0.053	0.900
	Water Body	0.081	0.012	0.145
Distance to River (m)	0–1953	0.214	0.294	1.372
	1953–4711	0.218	0.261	1.195
	4711–8044	0.218	0.172	0.786
	8044–12,576	0.215	0.166	0.773
	> 12,576	0.135	0.108	0.801
Groundwater Extraction (m³/day)	0–60	0.108	0.161	1.497
	60–180.15	0.314	0.339	1.081
	180.15–240.21	0.027	0.021	0.782
	241.21–330.28	0.272	0.192	0.706
	> 330.28	0.280	0.287	1.024
Distance to Fault (m)	0–946	0.212	0.252	1.188
	946–1972	0.207	0.213	1.031
	1972–3307	0.203	0.188	0.929
	3307–4339	0.199	0.191	0.960
	> 4339	0.180	0.156	0.868
Lithology	Qa	0.304	0.422	1.385
	PCEbgn	0.054	0.024	0.442
	PCEbngn	0.307	0.223	0.725
	PCEggn	0.011	0.007	0.609
	PCElbgn	0.003	0.000	0.000
	pgr	0.005	0.004	0.809
	Jsgr	0.045	0.068	1.518
	Jbgr	0.054	0.061	1.136
	PCEms	0.043	0.049	1.151
	PCEls	0.006	0.001	0.166
	Kkt	0.003	0.002	0.749
	rc	0.054	0.088	1.615
	PCEagn	0.010	0.002	0.199
	qz	0.001	0.000	0.228
	Qd	0.014	0.019	1.390
	Krh	0.004	0.004	0.992
	Qc	0.001	0.002	1.327
	mgn	0.003	0.000	0.000
	PCEpgn	0.009	0.002	0.215
	Jdgr	0.021	0.006	0.275
	PCEfgn	0.009	0.002	0.243
	PCEbs	0.005	0.002	0.423
	PCElgn	0.007	0.002	0.267
	Kct	0.006	0.004	0.686
	PCEqfgn	0.003	0.000	0.171
	PCEqf	0.003	0.000	0.000
	PCEsch	0.006	0.002	0.319
	PCEqgn	0.004	0.000	0.000
	Qr	0.004	0.004	0.873

Slope and aspect factor as a secondary feature of DEM may relate to the land subsidence occurrence because it can affect the soil infiltration in the landscape and water utility conditions [77,78]. The groundwater or sewage infiltration through a damaged pipe may erode the soil particles [77]. That condition can indirectly affect soil conditions that lead to land subsidence occurrence.

TWI is a secondary topographic variable that specifies the degree of water accumulation in a certain location; it is commonly used to quantify topographic influence on hydrological processes. The TWI of this study area was prepared based on the DEM and categorized into five classes. Profile curvature is a geomorphic property which shows the flow intensity, the amount of sediment, and erosion [79]. The profile curvature map was categorized into three classes: convex (less than −0.01), flat (−0.01–0.01), and concave (larger than 0.01).

Land-use is related to the ecological conditions and anthropological activities of land subsidence occurrence [80,81]. Variation in land use can explain the highly dissected zones within the region and provide insight into the land subsidence activity that is likely to occur. A land-use map was created based on a digital characteristics map provided by (National Geographic Information Institute) NGII; six land-use categories were analyzed in this study and the map is shown in Figure 3g. Underground water utilities are one of the most frequently cited factors impacting land subsidence [54,82].

Groundwater extraction can correlate to land subsidence event, especially in the underground structure [49]. Groundwater extraction map of this study area was prepared from annual average groundwater outflow data measured in 231 points from Seoul Government. Prior to spatial analysis, data on groundwater extraction should be converted into raster data. The accuracy of the raster map depends on the number of data points; however, the availability of groundwater data was limited in this study. To generate raster maps from these limited data, we applied the inverse distance weighted method to make statistical inferences using observed values before interpolating to create raster maps of groundwater extraction with a 30 m × 30 m cell size as shown in Figure 3h.

In order to evaluate the relationship between groundwater conditions and the occurrence of subsidence, several factors related to groundwater conditions can be evaluated such as distance to rivers [43,83]. The distance to rivers is represented by the proximity of the rivers and drainages in the study area [40]. The distance to the river map was calculated based on the map of the river provided by the National Geographic Information Institute (NGII). Then, buffers around the river were created (measured in meters), then the raster map was divided into five classes as shown in Figure 3i.

The geological parameters of a certain area may influence the occurrence of land subsidence, which is related to the lithological and structural variation which leads to differences in strength and permeability of rocks and soil. Lithology has also been an important feature to understand the land subsidence process by describing the structure of underground materials, as most cases of subsidence occurred on landfills and alluvium layers that have natural consolidation. Besides, the groundwater withdrawal and load from the building induce the compaction rate of the alluvium [17,62]. Any fluid present in the porous medium structure is under pressure because of the weight of the structure above it. If the fluid is withdrawn from below the surface, a decrease in pore pressure can occur, resulting in the loss of the supports and possibly lead to subsidence [74]. The lithology map can be seen in Figure 3j and the description of the lithology is shown in Table 4.

Table 4. Description of the lithological units in the study area.

Lithology ID	Description	Group
PCEagn	Granular gneiss	
PCEfgn	Fine granitic gneiss	
mgn	Hybrid gneiss	
PCElgn	White matter gneiss	Gneiss
PCElbgn	Lower arcuate gneiss	
PCEqfgn	Quartz feldspar gneiss	
PCEqf	Filigree gneiss	
PCEqgn	Quartz feldspar	
Jbgr	Biotite granite	
pgr	Geojeong Pyeonsang Granite	Granite
Jsgr	Selenite granite	
PCEbgn	Arctic black mica gneiss	
PCEbngn	Biotite Granite	Biotite Gneiss
PCEggn	Granitic gneiss	Granite Gneiss
Krh	Rhyolite	Rhyolite
Kkt	Lapiri tuff (mostly fused tuff)	Tuff
PCEls	Limestone	Limestone
Qa	Alluvium	Alluvium
PCEms	Mica schist	Mica schist
PCEsch	Gneiss schist	
PCEbs	Garnet black mica schist, ocular gneiss	Schist
rc	Red Sandstone, Conglomerate, Dark Red Conglomerate, Conglomerate.	Sedimentary Rock
Qd	Sand and clay	
Qc	Rock pieces, sand and clay	Sand and Clay
qz	Quartzite	quartz
Qr	Reclaimed land	Reclaimed land

Figure 3k shows a raster map of distance to the fault which used in this research. The presence of a fault line may weaken the porous medium structure and influence the subsidence occurrence, as in the case of Las Vegas, USA [84]. In this study, we used the fault lines as one of the factors of land subsidence to consider the impact of the fault line and the ground deformation. We performed a buffering distance from the fault line with the data published by the Korean Institute of Geoscience and Mineral Resources (KIGAM) with a 1:50,000 scale then categorized into five classes.

The spatial correlation for each factor was calculated using the frequency ratio and shown in Table 3. For the classes with frequency ratio values close to one or more, it shows a high correlation between subsidence and class of those factors, and vice versa [40]. From this calculation, it is considered that the subsidence has a correlation in the area with characteristics low elevation (0–30 m) and the flat area. The three classes of slope map (0–1.8, 1.8–3.86, 3.86–7.97 degree) and four classes of aspect map (Flat, North, East, Southeast) shows a spatial correlation with the land subsidence from this frequency ratio calculation. In addition, subsidence correlated with the drying area covered by building and non-permeable surface as shown in the land-use factor. Also, the ratio of alluvium (Qa) which dominates appear around the Han river exhibit a correlation with subsidence. Besides, there are six categories from this map that correlate with this calculation of lithology factor too. The land subsidence area has correlated with the fault distance in the area between 0–1972 m. The area with a groundwater extraction rate above 350 m^3/day has a spatial correlation with the land subsidence occurrences.

3. Results

The results from the PSI-StaMPS time-series analysis on deformation and the land-subsidence-susceptibility map are presented below. The time-series results were obtained by selecting location points at which subsidence occurred.

3.1. Land Subsidence Inventory Map

The land subsidence map from the Seoul Metropolitan Area was generated via the PSI-StaMPS method, using InSAR images in descending track captured from 2017 to 2020, and is shown in Figure 4a. To enhance measurement reliability, the vertical deformation map was generated using mean line-of-sight velocity [62]. Zooming locations for each selected point are shown in Figure 4b–e and the time-series graphs of all points are shown in Figure 4f,g.

In Figure 4b, represented the subsidence map in Gimpo the western part of this study area, with the black line indicating the subway line. In this area, especially in point A, there was a subsidence of 33.5 mm recorded with a mean deformation velocity of 12.57 mm/year from 2017 to 2020. The subsidence in Gimpo mostly occurs along the subway line that was newly operational in 2019. In this location, the subsidence is associated with the compressible deposits which consist of alluvium.

As can be seen from Figure 4c, the subsidence was exhibited around the subway line where the Shincheon subway station is located. Point B was recorded the maximum subsidence of up to 29 mm from 2017–2020 with the mean deformation velocity of 7.34 mm/year. This location consists of the intersection of subway line no 5 and subway line no 2, a residential area with high-density building also appears in this area. The subsidence in point B correlates with groundwater extraction and high-density building, as those conditions influence the subsidence rate in this area.

Figure 4d shows an overview of the subsidence near the Haengsin Station, known to be a depot for the metro train. The observation in point C revealed total subsidence of 34.35 mm from 2017 to 2020 and a mean deformation velocity of subsidence of 10.25 mm/year. The location is characterized by alluvium deposit that dominantly appears around the Han river. The geological features in this area show a correlation with the subsidence.

Figure 4e shows the subsidence map in Hanam city, the eastern part of the study area, with a black line indicates the subway line. The StaMPS result in point D shows the maximum subsidence of 26.17 mm from 2017 to 2020 with a mean deformation velocity rate of 8.42 mm/year. Hanam city is an area that has many developments such as residential areas and commercial buildings; subway construction is also being carried out in this area. Land subsidence can be related to underground work and building construction which pumping a large amount the groundwater [85]. In Hanam city, the subsidence is associated with urban land use and groundwater usage of this area.

Figure 4f,g show the time series graph from four selecting points in the study area. Generally, the periodic subsidence appeared in the vertical deformation graph. A possible reason for periodic subsidence in those areas was seasonal variation in the groundwater level and surface water loading. This result occurred due to the seasonal effect of groundwater extraction, where the selected points were surrounded by high-density buildings that mostly used groundwater as a water source. During the high season of groundwater withdrawal, the groundwater level decreased. After the rainy season, the groundwater level will rise and increase the aquifer system recovery (uplift) [84]. Those conditions may influence the deformation velocity in this study area.

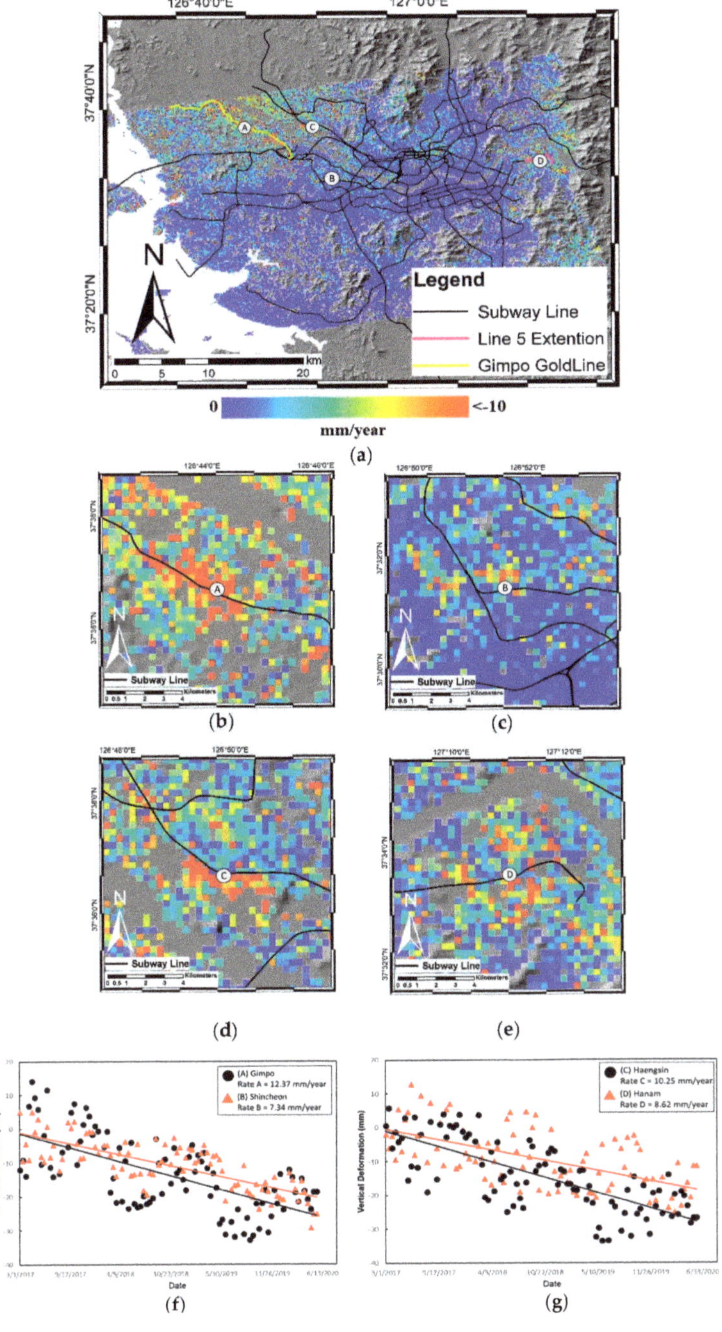

Figure 4. (**a**) Mean vertical deformation map in descending track of the Seoul Metropolitan Area generated from a Hillshade image comprising digital elevation model data from the Shuttle Radar Topography Mission. (**b**) Zooming of vertical deformation map at points A (Gimpo), (**c**) point B (Shincheon), (**d**) points C (Haengsin) and (**e**) points D (Hanam), (**f**) Vertical deformation time-series at points A (Gimpo) and B (Shincheon), and (**g**) points C (Haengsin) and D (Hanam).

3.2. Land Subsidence Susceptibility Map

Land subsidence susceptibility maps were constructed using the training dataset compiled from InSAR time-series data as land subsidence inventory map, ten land subsidence conditioning factors, and three different algorithms. Once the model training process was completed, susceptibility maps were constructed to visualize vulnerability to subsidence in the study area. In the land-subsidence-susceptibility map, each pixel in the study area was assigned a specific subsidence value using the quantile method [47].

Five susceptibility classes were used to reflect vulnerability to land subsidence: very low, low, moderate, high, and very high. Areas of very high susceptibility (marked red in Figure 5a–c) were most frequently found near the Han River and subway lines. The algorithms indicated that the northwestern area is very susceptible to subsidence, which may be due to several factors. For example, the geology of the northwestern area, which is near the Han River, is dominated by alluvium, which likely increases subsidence susceptibility. Most cases of observed subsidence have occurred on alluvium layers exhibiting natural consolidation; additionally, the increasing number of buildings and use of groundwater can exacerbate this condition [15,37,86]. A highly susceptible area was observed in the east, which may be associated with on-going construction in the same area. A few susceptible areas were observed along the northern Han River, mostly comprising high-density buildings and subway stations, but most of the northern area has low-to-moderate susceptibility to subsidence. Groundwater extraction in this area may increase the risk of subsidence, as some areas in which groundwater was extracted are now used for subway stations. Thus, the ground conditions in these areas might have been affected.

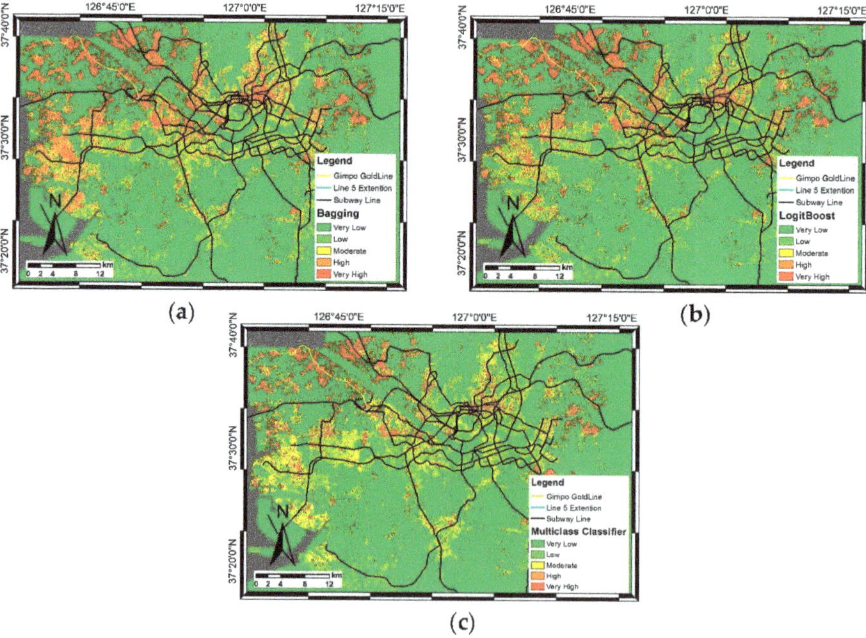

Figure 5. Land-subsidence-susceptibility map generated using three algorithms: the (**a**) Bagging, (**b**) LogitBoost, and (**c**) Multiclass Classifier algorithms.

Figure 6 shows the distribution of pixels in each susceptibility map generated by the meta-ensemble models. In the land-subsidence-susceptibility map generated using the Bagging model, 63.67% of the area exhibited very low susceptibility to subsidence, whereas 15.53%, 7.55%, 6.42%, and 7.04% of the area exhibited low, moderate, high, and very high susceptibility to subsidence, respectively. In the map constructed using the Multiclass Classifier model, 63.66% and 18.31% of the area exhibited very low and low susceptibility to subsidence, respectively, whereas 10.98%, 1.62%, and 5.42% of the area exhibited moderate, high, and very high susceptibility to subsidence, respectively. Lastly, based on the LogitBoost model, most of the area was not very susceptible to subsidence, with 64.44%, 18.07%, 6.62%, 4.41%, and 6.66% of the map classified as areas of very low, low, moderate, high, and very high susceptibility, respectively. The distribution of pixels in very low class and very high class in Figure 6 has a similar pattern between each algorithm. A very high class can be considered as the subsidence area. Meanwhile, medium and high classes are considered as areas of future land subsidence and very low and low classes are areas with the lowest probability of land subsidence in the future. With this description, it is possible to know the area and the extent of the potential for subsidence that will occur in the future. Generally, the consistency of this model can be evaluated based on the presence of past land subsidence in land subsidence susceptibility classes. The existence of a higher percentage of land subsidence pixels in a higher degree of susceptibility classes indicates higher consistency and vice-versa.

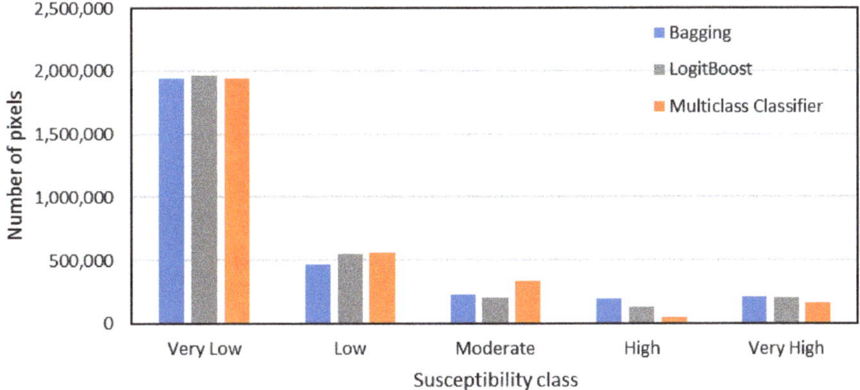

Figure 6. Distribution of pixels classified as areas of very low, low, moderate, high, and very high susceptibility in the land-subsidence-susceptibility maps generated by three machine-learning algorithms.

3.3. Model Validation

A good land-subsidence-susceptibility map should be able to predict future subsidence in the target area and provide initial information for preventative actions. To validate our susceptibility maps, the accuracy of all used algorithms in this study was evaluated by ROC curve analysis. ROC curve analysis has been used as a standard way of validating the probability models used to generate land subsidence susceptibility maps, according to the area under the curve (AUC) [6]. The AUC, which ranges from 0.5 to 1, was used to assess model accuracy. An AUC value near 0.5 indicates that a model is inaccurate, whereas a value near 1.0 indicates an ideal model with a good fit [50]. AUCs were calculated to compare model performance, with the model with the highest AUC value was taken to be the best model. The Bagging model produced the largest AUC (0.883), followed by the LogitBoost model (0.871) and the Multiclass Classifier model (0.856) as shown in Figure 7. Thus, the Bagging model generated the best subsidence-susceptibility map in this study. However, all models produced good AUC values, indicating that they all performed well in terms of predicting land-subsidence susceptibility in the study area.

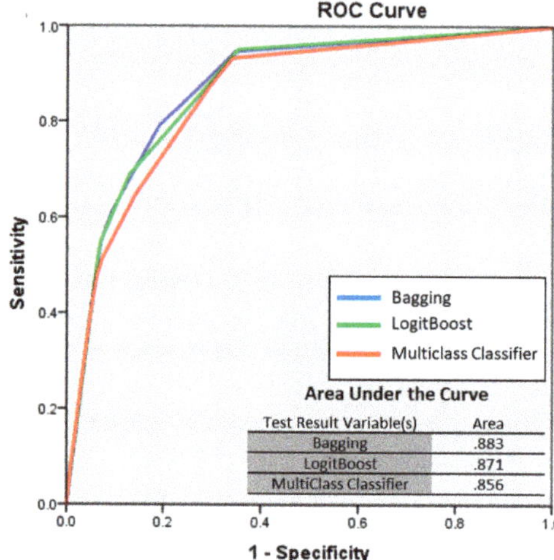

Figure 7. Receiver operating characteristic (ROC) curves associated with land-subsidence-susceptibility maps generated by three meta-ensemble algorithms (Bagging, LogitBoost, and Multiclass Classifier).

4. Discussion

4.1. Land Subsidence Inventory Map

StaMPS was employed to analyze land subsidence in the Seoul Metropolitan Area, with a deformation time-series map generated for all-terrain in the area based on descending-track data acquired from March 2017 to May 2020. Subsequently, a vertical deformation map was generated from the time-series analysis.

The results indicate that occurrences of subsidence were distributed over several locations in the study area, such as at Gimpo City (Figure 4a, point A). At this location, subsidence occurred near a new subway line that opened in 2019. Subsidence also occurred near a station with two subway lines near the southern Han River; high-density buildings are also found in this area (Figure 4a, point B). Other areas where subsidence occurred include a Seoul metro depot for several subway lines (Figure 4a, point C) and a newly developed area with several recently constructed residential and commercial buildings (Figure 4a, point D). Cases of land subsidence in the Seoul Metropolitan Area almost mostly occurred in the vicinity of subway lines or where the ground was weak. In particular, all areas of subsidence were located near subway lines and stations, implying that subway operation may be associated with subsidence in the study area [38,87]. Additionally, subway-tunnel excavations might have impacted the surrounding soil and the environment. During construction, how underground water is discharged, and the excavation method should be taken into consideration. Further analysis is needed to examine the impacts of construction on land subsidence in this area.

In terms of urbanization, the construction of buildings near subway lines and stations may add to the load on the soil and increase the risk of subsidence. High demand for transportation and urbanization increases the intensity of building construction and the amount of groundwater extracted. Besides, the groundwater extraction to fulfill social demand can influence the subsidence rate in the study area [3,88]. From the time-series deformation graph in Figure 4f, we can see a seasonal variation of subsidence rate in the study area between 2017–2020. It can also be noted that the high subsidence may appear in summer seasons. On the other hand, the subsidence rates lower after the rainy season which appears in July–August [3], the aquifer conditions after the rainy season are expected to have some influence on the subsidence rate. More study based on the water-level data analysis is required to better assess the possibility of the deformation, and further details of the structure and hydrologic parameters of groundwater should be resolved [89]. A combination of InSAR time-series analysis and analysis of hydrology of the area subsidence and the geomechanical parameters of the underlying aquifer structure area is a potential research topic to find out the cause of subsidence.

However, information extraction from the StaMPS technique is sometimes difficult due to a large number of PSs, thereby long interpretation times. The large amounts of PSs may cause several deficiencies in the analysis process, such as a reduction in the extraction of useful information from the dataset. Also, to obtain better time-series analysis results, several optimization steps can be taken for persistent scatterer points [20]. Several optimization methods for selecting PS points have been carried out [20,90]. This, in turn, could serve as a reference for future work on land subsidence studies that could increase efficiency and potentially lead to better deformation analyzes. Further studies should investigate other factors related to land subsidence. The results of GIS analysis are discussed below.

4.2. Land Subsidence Susceptibility Maps

Land subsidence should be mapped accurately to prepare subsidence inventories for target areas as an essential part of susceptibility analysis. To generate a subsidence-inventory map, we used InSAR remote-sensing data, which covered a broad area and were collected efficiently. Moreover, the InSAR time-series (StaMPS) method allows measuring of land subsidence and its rate to be measured to millimeter-level [37]. ArcGIS software was used for database construction, coordinate conversion, overlay analysis, and susceptibility modeling. Subsidence-related factors were identified based on information derived from the literature before susceptibility maps were generated [40,49]. Meta-ensemble machine learning was applied to estimate land-subsidence susceptibility, using three algorithms—Bagging, LogitBoost, and Multiclass Classifier.

Land subsidence susceptibility maps revealed that the northwestern and eastern areas, as well as a small area in the center, were most susceptible to land subsidence. We analyzed subsidence-related factors by comparing general patterns of subsidence with factor maps. The results revealed that most cases of subsidence occurred in areas where the ground consisted of alluvial layers, especially for subsidence that occurred near the Han River. However, there is a potential for subsidence in central areas that have different geological conditions from these two regions which are dominated by the alluvium layer. In this case, there are other factor influences besides geological factors in this subsidence modeling. The central area was assessed as moderately to highly susceptible to subsidence. In this area, there is a high density of buildings, and groundwater had been extracted at several spots near a subway station. Accordingly, groundwater outflow during subway operation could be another cause of land subsidence in this area [91]. If large amounts of groundwater are extracted, the surrounding soil structure may be affected. Thus, the weakened soil may be less able to withstand the pressure from aboveground buildings. Based on spatial distribution analysis, land use and groundwater extraction most strongly influence subsidence. In this study, the groundwater-extraction map was obtained via the interpolation of data on groundwater extraction near the subway infrastructure in Seoul, which might have generated errors in spatial distribution. Access to groundwater data for areas outside Seoul would allow for more accurate analyses of land subsidence.

In addition, several other factors that have not been identified in this study can be evaluated. The selection of these factors is based on the previous literature which may have some differences such as the condition of the area study and the subsidence mechanism. For this reason, additional analysis of factors related to the subsidence mechanism is needed to adjust to several concepts or assumptions of the subsidence mechanism that could potentially occur. However, adding some details such as aquifer conditions and groundwater levels can help evaluate the correlation of these factors and have the potential to improve the land subsidence susceptibility maps.

Susceptibility maps were validated based on ROC curves and AUC values and by comparing map predictions with testing data, which comprised 50% of the total dataset. The results indicate that the meta-ensemble approach performed better than the other approaches. A traditional model based on frequency ratio produced an AUC of 0.844, whereas the AUCs produced by the meta-ensemble models Multiclass Classifier, LogitBoost, and Bagging were 1.2%, 2.7%, and 2.95% larger, respectively.

Therefore, these techniques can reduce bias and account for factor weights to improve the accuracy of predictions.

Although all models produced good AUC values and thus performed well in terms of predicting land subsidence in the study area, the susceptibility map constructed using the Bagging model was the most accurate. These results agree well with previous findings that model performance in terms of predicting subsidence improves with the use of machine learning [92]. Therefore, the Bagging model should be used for the susceptibility map. In fact, the Bagging model uses more recently well-organized techniques in soft computing modeling that not only enable improvement of a single classifier but can also deal with complex and high-dimensional modeling problems. Given the complexity of land subsidence and the interaction of several related factors, novel combinations of model-method can considerably improve the accuracy of land subsidence prediction.

5. Conclusions

This study aimed to assess and map land-subsidence susceptibility in the Seoul Metropolitan Area using InSAR data from Sentinel-1 acquired between 2017 and 2020. A deformation time-series map was generated using StaMPS, which revealed that land subsidence occurred in four areas (Figure 4a), with subsidence rates of 6–12 mm/year. Subsidence mostly occurred near subway lines and where a new subway line was being constructed. Besides, the subsidence occurrence in areas with high-density building and heavy groundwater extraction may lead to weakening of the ground.

To identify the factors influencing land subsidence, 10 potential subsidence-related factors were analyzed. Factor maps were overlaid with subsidence maps and each pixel within a layer was evaluated in a GIS environment. The training and testing datasets were prepared from time-series InSAR from the Sentinel-1 SAR dataset using the StaMPS method. Then, the spatial correlation for each factor was calculated using the frequency ratio. Meta-ensemble algorithms (Bagging, LogitBoost, and Multiclass Classifier) were employed to generate land-subsidence-susceptibility maps, and model performance in terms of reliability and prediction accuracy was compared using ROC analysis.

The land-subsidence-susceptibility maps revealed that the northwestern and eastern areas, as well as a small central area, were most vulnerable to land subsidence. The susceptibility of the northwestern and eastern areas most appear in the geological condition which is dominated by alluvium. By contrast, in the central area, which is moderate to highly susceptible, land use and groundwater extraction are the main factors influencing subsidence risk. From the ROC analysis, the AUC produced by each model was computed. All models performed well (AUC > 0.8). Bagging produced the largest AUC of 0.883, followed by LogitBoost (0.871) and Multiclass Classifier (0.856). Compared with the frequency-ratio method, machine-learning models produced more accurate predictions and are thus more appropriate for subsidence analysis in this study area.

Accurate predictions are essential for environmental planning to control and mitigate the impacts of land subsidence. Land-subsidence-susceptibility mapping is a valuable method for identifying areas with a high risk of land subsidence. Despite limitations associated with the datasets used in this study, we demonstrated that the analysis of remote-sensing and GIS spatial data via the machine-learning approach generates reliable and accurate predictions of land subsidence. Further research is needed to determine the effect of aquifer conditions, subway construction and operation on land subsidence. A large dataset of PS points may influence a deficiency in extracting useful information. The optimization approaches for selecting PS points must be proposed to overcome those limitations in future work such as optimization hotspot analysis and other statistic methods [20,90]. Furthermore, with the high complexity of the relationship between land subsidence and other factors, a novel combination of a machine learning and meta-heuristic algorithm as a hybrid method can improve the results of the land subsidence susceptibility map.

Author Contributions: Conceptualization, C.-W.L.; methodology, C.-W.L., A.R.A., and M.F.F.; software, A.R.A. and M.F.F.; validation, C.-W.L. and M.F.F.; formal analysis, M.F.F.; investigations, C.-W.L.; resources, C.-W.L.; data curation, C.-W.L. and M.F.F.; writing—original draft preparation, M.F.F.; writing—review and editing, C.-W.L.,

A.R.A., and M.F.F.; visualization, C.-W.L. and M.F.F.; supervision, C.-W.L.; project administration, C.-W.L.; funding acquisition, C.-W.L. All authors have read and agreed to the published version of the manuscript.

Funding: This research was supported by a grant from the National Research Foundation of Korea (No. 2019R1A2C1085686), which is funded by the Government of Korea.

Conflicts of Interest: The authors declare no conflict of interest. The funders had no role in the design of the study; in the collection, analyses, or interpretation of data; in the writing of the manuscript; or in the decision to publish the results.

References

1. Machowski, R.; Rzetala, M.A.; Rzetala, M.; Solarski, M. Geomorphological and Hydrological Effects of Subsidence and Land use Change in Industrial and Urban Areas. *Land Degrad. Dev.* **2016**, *27*, 1740–1752. [CrossRef]
2. Jo, Y.-S.; Cho, S.-H.; Jang, Y.-S. Field investigation and analysis of ground sinking development in a metropolitan city, Seoul, Korea. *Environ. Earth Sci.* **2016**, *75*, 1353. [CrossRef]
3. Lee, H.; Oh, J. Establishing an ANN-based risk model for ground subsidence along railways. *Appl. Sci.* **2018**, *8*, 1936. [CrossRef]
4. Yuan, C.; Wang, X.; Wang, N.; Zhao, Q. Study on the Effect of Tunnel Excavation on Surface Subsidence Based on GIS Data Management. *Procedia Environ. Sci.* **2012**, *12*, 1387–1392. [CrossRef]
5. Roccheggiani, M.; Piacentini, D.; Tirincanti, E.; Perissin, D.; Menichetti, M. Detection and monitoring of tunneling induced ground movements using Sentinel-1 SAR interferometry. *Remote Sens.* **2019**, *11*, 639. [CrossRef]
6. Oh, H.J.; Lee, S. Assessment of ground subsidence using GIS and the weights-of-evidence model. *Eng. Geol.* **2010**, *115*, 36–48. [CrossRef]
7. Cigna, F.; Osmanoğlu, B.; Cabral-Cano, E.; Dixon, T.H.; Ávila-Olivera, J.A.; Garduño-Monroy, V.H.; DeMets, C.; Wdowinski, S. Monitoring land subsidence and its induced geological hazard with Synthetic Aperture Radar Interferometry: A case study in Morelia, Mexico. *Remote Sens. Environ.* **2012**, *117*, 146–161. [CrossRef]
8. Chen, B.; Gong, H.; Chen, Y.; Li, X.; Zhou, C.; Lei, K.; Zhu, L.; Duan, L.; Zhao, X. Land subsidence and its relation with groundwater aquifers in Beijing Plain of China. *Sci. Total Environ.* **2020**, *735*, 139111. [CrossRef]
9. Amelung, F.; Galloway, D.L.; Bell, J.W.; Zebker, H.A.; Laczniak, R.J. Sensing the ups and downs of Las Vegas: InSAR reveals structural control of land subsidence and aquifer-system deformation. *Geology* **1999**, *27*, 483–486. [CrossRef]
10. Huang, Y.; Jiang, X. Field-observed phenomena of seismic liquefaction and subsidence during the 2008 Wenchuan earthquake in China. *Nat. Hazards* **2010**, *54*, 839–850. [CrossRef]
11. Osmanoğlu, B.; Dixon, T.H.; Wdowinski, S.; Cabral-Cano, E.; Jiang, Y. Mexico City subsidence observed with persistent scatterer InSAR. *Int. J. Appl. Earth Obs. Geoinf.* **2011**, *13*, 1–12. [CrossRef]
12. Xu, Y.S.; Ma, L.; Shen, S.L.; Sun, W.J. Evaluation de la subsidence en considérant les structures constituant les aquifères de Shanghai, Chine. *Hydrogeol. J.* **2012**, *20*, 1623–1634. [CrossRef]
13. Ng, A.H.M.; Ge, L.; Li, X.; Abidin, H.Z.; Andreas, H.; Zhang, K. Mapping land subsidence in Jakarta, Indonesia using persistent scatterer interferometry (PSI) technique with ALOS PALSAR. *Int. J. Appl. Earth Obs. Geoinf.* **2012**, *18*, 232–242. [CrossRef]
14. Chaussard, E.; Amelung, F.; Abidin, H.; Hong, S.H. Sinking cities in Indonesia: ALOS PALSAR detects rapid subsidence due to groundwater and gas extraction. *Remote Sens. Environ.* **2013**, *128*, 150–161. [CrossRef]
15. Abidin, H.Z.; Djaja, R.; Darmawan, D.; Hadi, S.; Akbar, A.; Rajiyowiryono, H.; Sudibyo, Y.; Meilano, I.; Kasuma, M.A.; Kahar, J.; et al. Land subsidence of Jakarta (Indonesia) and its geodetic monitoring system. *Nat. Hazards* **2001**, *23*, 365–387. [CrossRef]
16. OECD. *Health at a Glance 2013: OECD Indicators*; OECD Publishing: Paris, France, 2013. [CrossRef]
17. Lee, J.Y.; Kwon, K.D.; Raza, M. Current water uses, related risks, and management options for Seoul megacity, Korea. *Environ. Earth Sci.* **2018**, *77*. [CrossRef]
18. Hanssen, R.F. *Radar Interferometry: Data Interpretation and Error Analysis*; Kluwer Academic: New York, NY, USA, 2010; ISBN 9789048156962.
19. Kang, Y.; Zhao, C.; Zhang, Q.; Lu, Z.; Li, B. Application of InSAR Techniques to an Analysis of the Guanling Landslide. *Remote Sens.* **2017**, *9*, 1046. [CrossRef]

20. Lu, P.; Bai, S.; Tofani, V.; Casagli, N. Landslides detection through optimized hot spot analysis on persistent scatterers and distributed scatterers. *ISPRS J. Photogramm. Remote Sens.* **2019**, *156*, 147–159. [CrossRef]

21. Perissin, D.; Ferretti, A. Urban-target recognition by means of repeated spaceborne SAR images. *IEEE Trans. Geosci. Remote Sens.* **2007**, *45*, 4043–4058. [CrossRef]

22. Motagh, M.; Shamshiri, R.; Haghshenas Haghighi, M.; Wetzel, H.U.; Akbari, B.; Nahavandchi, H.; Roessner, S.; Arabi, S. Quantifying groundwater exploitation induced subsidence in the Rafsanjan plain, southeastern Iran, using InSAR time-series and in situ measurements. *Eng. Geol.* **2017**, *218*, 134–151. [CrossRef]

23. Aly, M.H.; Klein, A.G.; Zebker, H.A.; Giardino, J.R. Land subsidence in the Nile Delta of Egypt observed by persistent scatterer interferometry. *Remote Sens. Lett.* **2012**, *3*, 621–630. [CrossRef]

24. Crosetto, M.; Monserrat, O.; Cuevas-González, M.; Devanthéry, N.; Crippa, B. Persistent Scatterer Interferometry: A review. *ISPRS J. Photogramm. Remote Sens.* **2016**, *115*, 78–89. [CrossRef]

25. Riddick, S.N.; Schmidt, D.A.; Deligne, N.I. An analysis of terrain properties and the location of surface scatterers from persistent scatterer interferometry. *ISPRS J. Photogramm. Remote Sens.* **2012**, *73*, 50–57. [CrossRef]

26. Khorrami, M.; Abrishami, S.; Maghsoudi, Y.; Alizadeh, B.; Perissin, D. Extreme subsidence in a populated city (Mashhad) detected by PSInSAR considering groundwater withdrawal and geotechnical properties. *Sci. Rep.* **2020**, *10*, 1–16. [CrossRef] [PubMed]

27. Yazici, B.V.; Tunc Gormus, E. Investigating persistent scatterer InSAR (PSInSAR) technique efficiency for landslides mapping: A case study in Artvin dam area, in Turkey. *Geocarto Int.* **2020**, 1–19. [CrossRef]

28. Jiaxuan, H.; Mowen, X.; Atkinson, P.M. Dynamic susceptibility mapping of slow-moving landslides using PSInSAR. *Int. J. Remote Sens.* **2020**, *41*, 7509–7529. [CrossRef]

29. Tessitore, S.; Fernández-Merodo, J.A.; Herrera, G.; Tomás, R.; Ramondini, M.; Sanabria, M.; Duro, J.; Mulas, J.; Calcaterra, D. Comparison of water-level, extensometric, DInSAR and simulation data for quantification of subsidence in Murcia City (SE Spain). *Hydrogeol. J.* **2016**, *24*, 727–747. [CrossRef]

30. Wang, H.; Feng, G.; Xu, B.; Yu, Y.; Li, Z.; Du, Y.; Zhu, J. Deriving spatio-temporal development of ground subsidence due to subway construction and operation in Delta regions with PS-InSAR data: A case study in Guangzhou, China. *Remote Sens.* **2017**, *9*, 1004. [CrossRef]

31. Khakim, M.Y.N.; Tsuji, T.; Matsuoka, T. Lithology-controlled subsidence and seasonal aquifer response in the Bandung basin, Indonesia, observed by synthetic aperture radar interferometry. *Int. J. Appl. Earth Obs. Geoinf.* **2014**, *32*, 199–207. [CrossRef]

32. Zhou, C.; Gong, H.; Chen, B.; Gao, M.; Cao, Q.; Cao, J.; Duan, L.; Zuo, J.; Shi, M. Land Subsidence Response to Different Land Use Types and Water Resource Utilization in Beijing-Tianjin-Hebei, China. *Remote Sens.* **2020**, *12*, 457. [CrossRef]

33. Lu, P.; Han, J.; Hao, T.; Li, R.; Qiao, G. Seasonal Deformation of Permafrost in Wudaoliang Basin in Qinghai-Tibet Plateau Revealed by StaMPS-InSAR. *Mar. Geod.* **2020**, *43*, 248–268. [CrossRef]

34. Jennifer, J.J.; Saravanan, S.; Pradhan, B. Persistent Scatterer Interferometry in the post-event monitoring of the Idukki Landslides. *Geocarto Int.* **2020**. [CrossRef]

35. Tzampoglou, P.; Loupasakis, C. Mining geohazards susceptibility and risk mapping: The case of the Amyntaio open-pit coal mine, West Macedonia, Greece. *Environ. Earth Sci.* **2017**, *76*. [CrossRef]

36. Tofani, V.; Raspini, F.; Catani, F.; Casagli, N. Persistent Scatterer Interferometry (PSI) Technique for Landslide Characterization and Monitoring. *Remote Sens.* **2013**, *5*, 1045–1065. [CrossRef]

37. Bianchini, S.; Solari, L.; Del Soldato, M.; Raspini, F.; Montalti, R.; Ciampalini, A.; Casagli, N. Ground Subsidence Susceptibility (GSS) Mapping in Grosseto Plain (Tuscany, Italy) Based on Satellite InSAR Data Using Frequency Ratio and Fuzzy Logic. *Remote Sens.* **2019**, *11*, 2015. [CrossRef]

38. Ng, A.H.M.; Wang, H.; Dai, Y.; Pagli, C.; Chen, W.; Ge, L.; Du, Z.; Zhang, K. InSAR reveals land deformation at Guangzhou and Foshan, China between 2011 and 2017 with COSMO-SkyMed data. *Remote Sens.* **2018**, *10*, 813. [CrossRef]

39. Lan, H.X.; Zhou, C.H.; Wang, L.J.; Zhang, H.Y.; Li, R.H. Landslide hazard spatial analysis and prediction using GIS in the Xiaojiang watershed, Yunnan, China. *Eng. Geol.* **2004**, *76*, 109–128. [CrossRef]

40. Pradhan, B.; Abokharima, M.H.; Jebur, M.N.; Tehrany, M.S. Land subsidence susceptibility mapping at Kinta Valley (Malaysia) using the evidential belief function model in GIS. *Nat. Hazards* **2014**, *73*, 1019–1042. [CrossRef]

41. Regmi, A.D.; Yoshida, K.; Nagata, H.; Pradhan, A.M.S.; Pradhan, B.; Pourghasemi, H.R. The relationship between geology and rock weathering on the rock instability along Mugling-Narayanghat road corridor, Central Nepal Himalaya. *Nat. Hazards* **2013**, *66*, 501–532. [CrossRef]

42. Lee, S.; Park, I.; Choi, J.K. Spatial prediction of ground subsidence susceptibility using an artificial neural network. *Environ. Manag.* **2012**, *49*, 347–358. [CrossRef]

43. Arabameri, A.; Saha, S.; Roy, J.; Tiefenbacher, J.P.; Cerda, A.; Biggs, T.; Pradhan, B.; Thi Ngo, P.T.; Collins, A.L. A novel ensemble computational intelligence approach for the spatial prediction of land subsidence susceptibility. *Sci. Total Environ.* **2020**, *726*, 138595. [CrossRef] [PubMed]

44. Pradhan, B. A comparative study on the predictive ability of the decision tree, support vector machine and neuro-fuzzy models in landslide susceptibility mapping using GIS. *Comput. Geosci.* **2013**, *51*, 350–365. [CrossRef]

45. Wu, W.; Zucca, C.; Karam, F.; Liu, G. Enhancing the performance of regional land cover mapping. *Int. J. Appl. Earth Obs. Geoinf.* **2016**, *52*, 422–432. [CrossRef]

46. Tien Bui, D.; Ho, T.C.; Revhaug, I.; Pradhan, B.; Nguyen, D.B. Landslide Susceptibility Mapping Along the National Road 32 of Vietnam Using GIS-Based J48 Decision Tree Classifier and Its Ensembles. In *Cartography from Pole to Pole*; Springer: Berlin/Heidelberg, Germany, 2014; pp. 303–317.

47. Kadavi, P.R.; Lee, C.W.; Lee, S. Application of ensemble-based machine learning models to landslide susceptibility mapping. *Remote Sens.* **2018**, *10*, 1252. [CrossRef]

48. Hong, H.; Liu, J.; Zhu, A.X. Modeling landslide susceptibility using LogitBoost alternating decision trees and forest by penalizing attributes with the bagging ensemble. *Sci. Total Environ.* **2020**, *718*, 137231. [CrossRef] [PubMed]

49. Abdollahi, S.; Pourghasemi, H.R.; Ghanbarian, G.A.; Safaeian, R. Prioritization of effective factors in the occurrence of land subsidence and its susceptibility mapping using an SVM model and their different kernel functions. *Bull. Eng. Geol. Environ.* **2019**, *78*, 4017–4034. [CrossRef]

50. Fawcett, T. An introduction to ROC analysis. *Pattern Recognit. Lett.* **2006**, *27*, 861–874. [CrossRef]

51. Kim, Y.Y.; Lee, K.K.; Sung, I.H. Urbanization and the groundwater budget, metropolitan Seoul area, Korea. *Hydrogeol. J.* **2001**, *9*, 401–412. [CrossRef]

52. Korea, S. *Complete Enumeration Results of the 2010 Population and Housing Census*; Statistics Korea: Daejeon, Korea, 2011.

53. Choi, B.Y.; Yun, S.T.; Yu, S.Y.; Lee, P.K.; Park, S.S.; Chae, G.T.; Mayer, B. Hydrochemistry of urban groundwater in Seoul, South Korea: Effects of land-use and pollutant recharge. *Environ. Geol.* **2005**, *48*, 979–990. [CrossRef]

54. Chae, G.T.; Yun, S.T.; Kim, D.S.; Kim, K.H.; Joo, Y. Time-series analysis of three years of groundwater level data (Seoul, South Korea) to characterize urban groundwater recharge. *Q. J. Eng. Geol. Hydrogeol.* **2010**, *43*, 117–127. [CrossRef]

55. Korea, R. *KORAIL Sustainability Report 2015*; Korea Railroad Corporation: Daejeon, Korea, 2015.

56. Farr, T.G.; Rosen, P.A.; Caro, E.; Crippen, R.; Duren, R.; Hensley, S.; Kobrick, M.; Paller, M.; Rodriguez, E.; Roth, L.; et al. The shuttle radar topography mission. *Rev. Geophys.* **2007**, *45*. [CrossRef]

57. Hooper, A.; Segall, P.; Zebker, H. Persistent scatterer interferometric synthetic aperture radar for crustal deformation analysis, with application to Volcán Alcedo, Galápagos. *J. Geophys. Res. Solid Earth* **2007**, *112*. [CrossRef]

58. Hooper, A.J. A multi-temporal InSAR method incorporating both persistent scatterer and small baseline approaches. *Geophys. Res. Lett.* **2008**, *35*. [CrossRef]

59. Sousa, J.J.; Hooper, A.J.; Hanssen, R.F.; Bastos, L.C.; Ruiz, A.M. Persistent Scatterer InSAR: A comparison of methodologies based on a model of temporal deformation vs. spatial correlation selection criteria. *Remote Sens. Environ.* **2011**, *115*, 2652–2663. [CrossRef]

60. Pepe, A.; Bonano, M.; Zhao, Q.; Yang, T.; Wang, H. The Use of C-/X-Band Time-Gapped SAR Data and Geotechnical Models for the Study of Shanghai's Ocean-Reclaimed Lands through the SBAS-DInSAR Technique. *Remote Sens.* **2016**, *8*, 911. [CrossRef]

61. Ren, H.; Feng, X. Calculating vertical deformation using a single InSAR pair based on singular value decomposition in mining areas. *Int. J. Appl. Earth Obs. Geoinf.* **2020**, *92*, 102115. [CrossRef]

62. Yastika, P.; Shimizu, N.; Abidin, H.Z. Monitoring of long-term land subsidence from 2003 to 2017 in coastal area of Semarang, Indonesia by SBAS DInSAR analyses using Envisat-ASAR, ALOS-PALSAR, and Sentinel-1A SAR data. *Adv. Space Res.* **2019**, *63*, 1719–1736. [CrossRef]

63. Lee, S.; Park, I. Application of decision tree model for the ground subsidence hazard mapping near abandoned underground coal mines. *J. Environ. Manag.* **2013**, *127*, 166–176. [CrossRef]

64. Korjus, K.; Hebart, M.N.; Vicente, R. An Efficient Data Partitioning to Improve Classification Performance While Keeping Parameters Interpretable. *PLoS ONE* **2016**, *11*, e0161788. [CrossRef]

65. Jaafari, A.; Zenner, E.K.; Panahi, M.; Shahabi, H. Hybrid artificial intelligence models based on a neuro-fuzzy system and metaheuristic optimization algorithms for spatial prediction of wildfire probability. *Agric. For. Meteorol.* **2019**, *266–267*, 198–207. [CrossRef]

66. Kim, K.D.; Lee, S.; Oh, H.J.; Choi, J.K.; Won, J.S. Assessment of ground subsidence hazard near an abandoned underground coal mine using GIS. *Environ. Geol.* **2006**, *50*, 1183–1191. [CrossRef]

67. Silalahi, F.E.S.; Arifianti, Y.; Hidayat, F. Landslide susceptibility assessment using frequency ratio model in Bogor, West Java, Indonesia. *Geosci. Lett.* **2019**, *6*, 1–17. [CrossRef]

68. Breiman, L. Bagging predictors. *Mach. Learn.* **1996**, *24*, 123–140. [CrossRef]

69. Sedano, J.; Gonzalez, S.; Herrero, A.; Baruque, B.; Corchado, E. Mutating network scans for the assessment of supervised classifier ensembles. *Log. J. IGPL* **2013**, *21*, 630–647. [CrossRef]

70. Friedman, J.; Hastie, T.; Tibshirani, R. Additive logistic regression: A statistical view of boosting. *Ann. Stat.* **2000**, *28*, 337–407. [CrossRef]

71. Pourghasemi, H.; Gayen, A.; Park, S.; Lee, C.-W.; Lee, S. Assessment of Landslide-Prone Areas and Their Zonation Using Logistic Regression, LogitBoost, and NaïveBayes Machine-Learning Algorithms. *Sustainability* **2018**, *10*, 3697. [CrossRef]

72. Cai, Y.D.; Feng, K.Y.; Lu, W.C.; Chou, K.C. Using LogitBoost classifier to predict protein structural classes. *J. Theor. Biol.* **2006**, *238*, 172–176. [CrossRef] [PubMed]

73. Kowsari, K.; Brown, D.E.; Heidarysafa, M.; Meimandi, K.J.; Gerber, M.S.; Barnes, L.E. HDLTex: Hierarchical Deep Learning for Text Classification. In Proceedings of the 16th IEEE International Conference on Machine Learning and Applications, ICMLA 2017, Cancun, Mexico, 18–21 December 2017; Institute of Electrical and Electronics Engineers Inc.: Piscataway, NJ, USA, 2017; pp. 364–371.

74. Gambolati, G.; Teatini, P. Geomechanics of subsurface water withdrawal and injection. *Water Resour. Res.* **2015**, *51*, 3922–3955. [CrossRef]

75. Erban, L.E.; Gorelick, S.M.; Zebker, H.A. Groundwater extraction, land subsidence, and sea-level rise in the Mekong Delta, Vietnam. *Environ. Res. Lett.* **2014**, *9*, 084010. [CrossRef]

76. Suh, J.; Choi, Y.; Park, H.D. GIS-based evaluation of mining-induced subsidence susceptibility considering 3D multiple mine drifts and estimated mined panels. *Environ. Earth Sci.* **2016**, *75*, 1–19. [CrossRef]

77. Kim, K.; Kim, J.; Kwak, T.Y.; Chung, C.K. Logistic regression model for sinkhole susceptibility due to damaged sewer pipes. *Nat. Hazards* **2018**, *93*, 765–785. [CrossRef]

78. Del Giudice, G.; Padulano, R.; Siciliano, D. Multivariate probability distribution for sewer system vulnerability assessment under data-limited conditions. *Water Sci. Technol.* **2016**, *73*, 751–760. [CrossRef] [PubMed]

79. Yesilnacar, E.; Topal, T. Landslide susceptibility mapping: A comparison of logistic regression and neural networks methods in a medium scale study, Hendek region (Turkey). *Eng. Geol.* **2005**, *79*, 251–266. [CrossRef]

80. Andaryani, S.; Nourani, V.; Trolle, D.; Dehgani, M.; Asl, A.M. Assessment of land use and climate change effects on land subsidence using a hydrological model and radar technique. *J. Hydrol.* **2019**, *578*, 124070. [CrossRef]

81. Minderhoud, P.S.J.; Coumou, L.; Erban, L.E.; Middelkoop, H.; Stouthamer, E.; Addink, E.A. The relation between land use and subsidence in the Vietnamese Mekong delta. *Sci. Total Environ.* **2018**, *634*, 715–726. [CrossRef]

82. Yoo, C. Ground settlement during tunneling in groundwater drawdown environment—Influencing factors. *Undergr. Space* **2016**, *1*, 20–29. [CrossRef]

83. Arabameri, A.; Lee, S.; Tiefenbacher, J.P.; Ngo, P.T.T. Novel Ensemble of MCDM-Artificial Intelligence Techniques for Groundwater-Potential Mapping in Arid and Semi-Arid Regions (Iran). *Remote Sens.* **2020**, *12*, 490. [CrossRef]

84. Bell, J.W.; Amelung, F.; Ferretti, A.; Bianchi, M.; Novali, F. Permanent scatterer InSAR reveals seasonal and long-term aquifer-system response to groundwater pumping and artificial recharge. *Water Resour. Res.* **2008**, *44*. [CrossRef]

85. Lee, J.Y. Lessons from three groundwater disputes in Korea: Lack of comprehensive and integrated investigation. *Int. J. Water* **2017**, *11*, 59–72. [CrossRef]

86. Notti, D.; Mateos, R.M.; Monserrat, O.; Devanthéry, N.; Peinado, T.; Roldán, F.J.; Fernández-Chacón, F.; Galve, J.P.; Lamas, F.; Azañón, J.M. Lithological control of land subsidence induced by groundwater withdrawal in new urban AREAS (Granada Basin, SE Spain). Multiband DInSAR monitoring. *Hydrol. Process.* **2016**, *30*, 2317–2331. [CrossRef]

87. Chen, W.F.; Gong, H.L.; Chen, B.B.; Liu, K.S.; Gao, M.; Zhou, C.F. Spatiotemporal evolution of land subsidence around a subway using InSAR time-series and the entropy method. *GISci. Remote Sens.* **2017**, *54*, 78–94. [CrossRef]

88. Reeves, J.A.; Knight, R.; Zebker, H.A.; Kitanidis, P.K.; Schreüder, W.A. Estimating temporal changes in hydraulic head using InSAR data in the San Luis Valley, Colorado. *Water Resour. Res.* **2014**, *50*, 4459–4473. [CrossRef]

89. Motagh, M.; Walter, T.R.; Sharifi, M.A.; Fielding, E.; Schenk, A.; Anderssohn, J.; Zschau, J. Land subsidence in Iran caused by widespread water reservoir overexploitation. *Geophys. Res. Lett.* **2008**, *35*. [CrossRef]

90. Lu, P.; Casagli, N.; Catani, F.; Tofani, V. Persistent scatterers interferometry hotspot and cluster analysis (PSI-HCA) for detection of extremely slow-moving landslides. *Int. J. Remote Sens.* **2012**, *33*, 466–489. [CrossRef]

91. Seong, J.-H. *The Contiguity Ground and Structures Sinkage Analysis of in City Excavation*; Korean Geotechnical Society: Seoul, Korea, 2009.

92. Rahmati, O.; Golkarian, A.; Biggs, T.; Keesstra, S.; Mohammadi, F.; Daliakopoulos, I.N. Land subsidence hazard modeling: Machine learning to identify predictors and the role of human activities. *J. Environ. Manag.* **2019**, *236*, 466–480. [CrossRef]

Publisher's Note: MDPI stays neutral with regard to jurisdictional claims in published maps and institutional affiliations.

remote sensing

MDPI

Article

Mapping Urban Green Spaces at the Metropolitan Level Using Very High Resolution Satellite Imagery and Deep Learning Techniques for Semantic Segmentation

Roberto E. Huerta [1], Fabiola D. Yépez [1,*], Diego F. Lozano-García [2], Víctor H. Guerra Cobián [1], Adrián L. Ferriño Fierro [1], Héctor de León Gómez [1], Ricardo A. Cavazos González [1] and Adriana Vargas-Martínez [2]

1   Facultad de Ingeniería Civil, Universidad Autónoma de Nuevo León, San Nicolás de los Garza 66455, Mexico; roberto.huertagc@uanl.edu.mx (R.E.H.); victor.guerracb@uanl.edu.mx (V.H.G.C.); adrian.ferrinofr@uanl.edu.mx (A.L.F.F.); hector.deleongm@uanl.edu.mx (H.d.L.G.); ricardo.cavazosgzz@uanl.edu.mx (R.A.C.G.)
2   Escuela de Ingeniería y Ciencias, Tecnológico de Monterrey, Ave. Eugenio Garza Sada 2501, Monterrey 64849, Mexico; dflozano@tec.mx (D.F.L.-G.); adriana.vargas.mtz@tec.mx (A.V.M.)
*   Correspondence: fabiola.yepezrn@uanl.edu.mx; Tel.: +52-8114-424-4000 (ext. 7258)

Citation: Huerta, R.E.; Yépez, F.D.; Lozano-García, D.F.; Guerra Cobián, V.H.; Ferriño Fierro, A.L.; de León Gómez, H.; Cavazos González, R.A.; Vargas-Martínez, A. Mapping Urban Green Spaces at the Metropolitan Level Using Very High Resolution Satellite Imagery and Deep Learning Techniques for Semantic Segmentation. *Remote Sens.* **2021**, *13*, 2031. https://doi.org/10.3390/rs13112031

Academic Editors: Chang-Wook Lee, Hyangsun Han, Hoonyol Lee and Yu-Chul Park

Received: 10 March 2021
Accepted: 27 April 2021
Published: 21 May 2021

Publisher's Note: MDPI stays neutral with regard to jurisdictional claims in published maps and institutional affiliations.

Abstract: Urban green spaces (UGSs) provide essential environmental services for the well-being of ecosystems and society. Due to the constant environmental, social, and economic transformations of cities, UGSs pose new challenges for management, particularly in fast-growing metropolitan areas. With technological advancement and the evolution of deep learning, it is possible to optimize the acquisition of UGS inventories through the detection of geometric patterns present in satellite imagery. This research evaluates two deep learning model techniques for semantic segmentation of UGS polygons with the use of different convolutional neural network encoders on the U-Net architecture and very high resolution (VHR) imagery to obtain updated information on UGS polygons at the metropolitan area level. The best model yielded a Dice coefficient of 0.57, IoU of 0.75, recall of 0.80, and kappa coefficient of 0.94 with an overall accuracy of 0.97, which reflects a reliable performance of the network in detecting patterns that make up the varied geometry of UGSs. A complete database of UGS polygons was quantified and categorized by types with location and delimited by municipality, allowing for the standardization of the information at the metropolitan level, which will be useful for comparative analysis with a homogenized and updated database. This is of particular interest to urban planners and UGS decision-makers.

Keywords: neural networks; urban vegetation; urban open spaces; Monterrey Metropolitan Area; sustainable development

1. Introduction

Urban green spaces (UGSs) face significant challenges due to rapid urbanization and climate change [1]. UGSs are crucial in order to safeguard the quality of urban life [2]. City managers are urged to integrate UGSs in urban development plans [3]. The conservation of ecosystem services of UGSs can mitigate the impacts of urban development; can reduce ecological debts; and is the simplest, fastest, and most effective way to ameliorate important challenges in cities, such as urban heat islands and air pollution [4–6]. UGSs enhance resilience, health, and quality of life of citizens, especially benefiting those with high accessibility. UGS accessibility is a crucial aspect of sustainable urban planning [7] and social justice [8].

UGS survey data are not commonly updated or freely accessible to local users. There is a need for uniform and spatially explicit inventories of existing UGSs [9] and the quantification of their proximity services [10]. One of the most critical functions of the UGSs within cities is the provision of essential environmental services, as it is related to human well-being [11,12]. UGSs contribute to the reduction of harmful effects that cause cardiovascular, respiratory, and metabolic diseases [13], and they mitigate the stress caused by increases in temperature and noise levels [14]. Additionally, UGSs promote physical activity and social interaction, improving the physical and mental health of residents who use these facilities [15].

The constant environmental, social, and economic transformations of cities make the management of UGSs a major challenge for government administrations in metropolitan areas with extensive and rapid urban development [16,17]. Depending on the management of the UGSs, negative or positive effects can be promoted, from the destruction of these spaces [18] to promoting adequate conditions for management and maintenance [19]. Inventories of UGSs allow the monitoring of their status and help provide guidelines for the development of adequate management strategies [20].

UGSs exist in diverse shapes, sizes, vegetation covers, and types (i.e., park, residential garden, median strip, square, and roundabout) [21,22]. Traditional methods to obtain polygons of UGSs have relied on the visual interpretation of aerial imagery, remotely sensed data interpretation, and manual digitalization [23]. Similar to tree inventories and other natural elements in UGSs, data collection methods are intensive and involve manual measurements of dendometric parameters in the field. These methods are time-consuming and costly [24]. The integration of remote sensing and geographic information systems for mapping and monitoring UGSs has been advantageous, as this reduces the resources required by traditional methods [25–27]. Moreover, the use of these techniques and frameworks allows the computation of vegetation indices that highlight vegetation properties such as vegetation cover and vigor [28].

Methods based on different machine learning algorithms, including decision tree [29], maximum likelihood classification [30], random forest, and support vector regression [31], have been used to map UGSs. Other authors propose the use of object-based image analysis, which takes advantage of both the spectral and contextual information of the classifying objects [32]. With technological advancement and the evolution of deep learning, optimization of the acquisition of UGS inventories is possible through the detection of spectral and geometric patterns available in satellite imagery [33]. Convolutional neural networks (CNNs) have performed well at high-level vision tasks, such as image classification, object detection, and semantic segmentation [34]. A combination of multitemporal MODIS and Landsat-7 imagery was used to classify UGSs in Mumbai metropolitan area in India [35]. The results indicate that for over 15 years, the overall UGSs were reduced to 50%. Other authors analyzed four different methods of classifying UGSs: support vector machine, random forest, artificial neural networks, and naïve Bayes classifier [36]. They found that support vector machines produce higher accuracy classifications in a short amount of time. Multitemporal high-resolution imagery was employed to map open spaces in Kampala, Uganda, with the use of a cloud computation method and machine learning that combined nine base classifiers [37]. The results produced a map of open spaces with an 88% classification accuracy. A deep learning classification based upon a high-resolution network (HRNet) method of high-resolution GaoFen-2 imagery was used for the city of Beijing, China, indicating that the HRNet combined with phenological analysis significantly improved the classification of UGSs [38].

CNNs are convenient models for semantic segmentation because they produce hierarchies that help determine low-, medium-, and high-level characteristics [39,40]. These models are automatically trained using previously labeled input information, and they produce class identification results [41]. With the use of labeled samples, a network can update its weights until it obtains a proper mapping of the inputs and a minimal loss [42].

Due to the absence of a dense layer, the use of fully convolutional networks (FCNs) allows the generation of outputs in which each pixel has a classification according to the input information [43]. Based on the FCN model, the U-Net architecture uses the same principle and considers a symmetric encoder–decoder composition. This process first reduces the size and increases the number of bands of the training images and their activation maps generated in each layer of the network to subsequently carry out the opposite process considering information from the encoder in the segmentation of fine details [44]. These types of networks have achieved wide success with state-of-the-art results for a wide variety of problems from medical applications [45,46] to their employment in remote sensing for road [47] and building extractions [48], as well as land cover classification [49], but they have not been used to make many advances in the UGS area.

Detailed geometric information on UGSs is typically presented as a shapefile that is not updated frequently; therefore, it does not reflect changes occurring due to rapid urban development processes. Additionally, spatial data or information about the availability of UGSs is not generally accessible to urban residents [50], restricting their use. The need to improve and make available geospatial data of green and public spaces is recognized by the United Nations sustainable development agenda as it helps to create more inclusive, safe, resilient, sustainable cities [51]. The generation of UGS inventories in conjunction with other public space inventories aid in the calculation of Sustainable Development Goal (SDG) 11.7.1, i.e., quantifying the average share of green and public spaces in cities. This allows for the obtainment of SGD 11.7, which is to "provide universal access to safe, inclusive and accessible, green and public spaces". It is therefore necessary to later relate the information available in digitization, vectorization, and computation to demographic data to generate accessibility maps [52,53]. This study evaluates two deep learning model techniques for semantic segmentation of UGS polygons. The process involves different convolutional neural network encoders on the U–Net architecture with the use of three-band compositions of very high resolution (VHR) satellite imagery channels and vegetation indices as input data. This precise and updated data collection and new UGS cartography at the metropolitan level would improve the understanding of connectivity and accessibility of UGSs as a basis for management and decision-making for land use in urban areas.

2. Materials and Methods

2.1. Study Area

The study area chosen to test this method was the Monterrey Metropolitan Area (MMA) (Figure 1), located at the coordinates 25°40′00″ N 100°18′00″ W. It has a total area of 6687.10 km² of which 27.57% is built-up area. The MMA is comprised of Monterrey, the capital of the State of Nuevo Leon, and 11 surrounding municipalities [54]. Its population, as of 2015, was 4.7 million inhabitants [55]. Within its orography, the Sierra Madre Oriental, Sierra San Miguel, the hills of Topo Chico, La Silla, and Las Mitras are prominent.

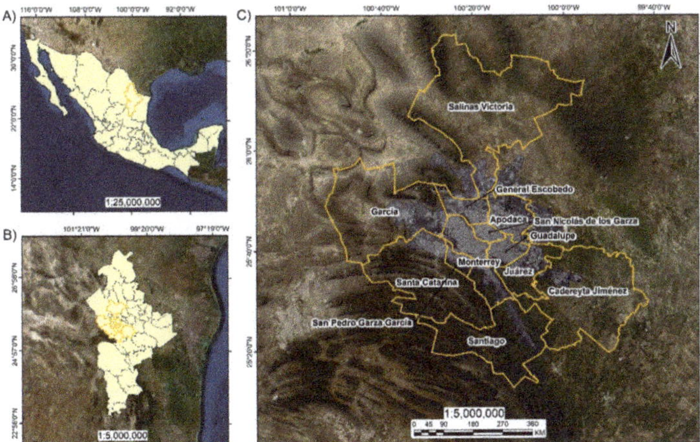

Figure 1. Study area location. (**A**) State of Nuevo Leon within the Mexican Republic; (**B**) Monterrey Metropolitan Area (MMA) within the state of Nuevo Leon; (**C**) orthomosaic of WorldView-2 coverage of the MMA.

The methodological workflow is shown in Figure 2, where the methodology is divided into three sections: data preprocessing, CNN model implementation, and evaluation of semantic segmentation of UGSs. For this study, we used a UGS definition based on the "Regulation of Environmental Protection and Urban Aesthetics of Monterrey" [56]. This document describes UGSs as land surfaces containing vegetation, gardens, groves, and complementary minor buildings for public use within the urban area or its periphery. Input label polygons for the CNN models were obtained from three sources, the UGS database of the National Geostatistical Framework (2010) of the National Institute of Statistics and Geography (INEGI), the collaborative Open Street Maps (OSM) project [57], and the database of median strips of the MMA arranged by the Department of Geomatics of the Institute of Civil Engineering of the UANL [58].

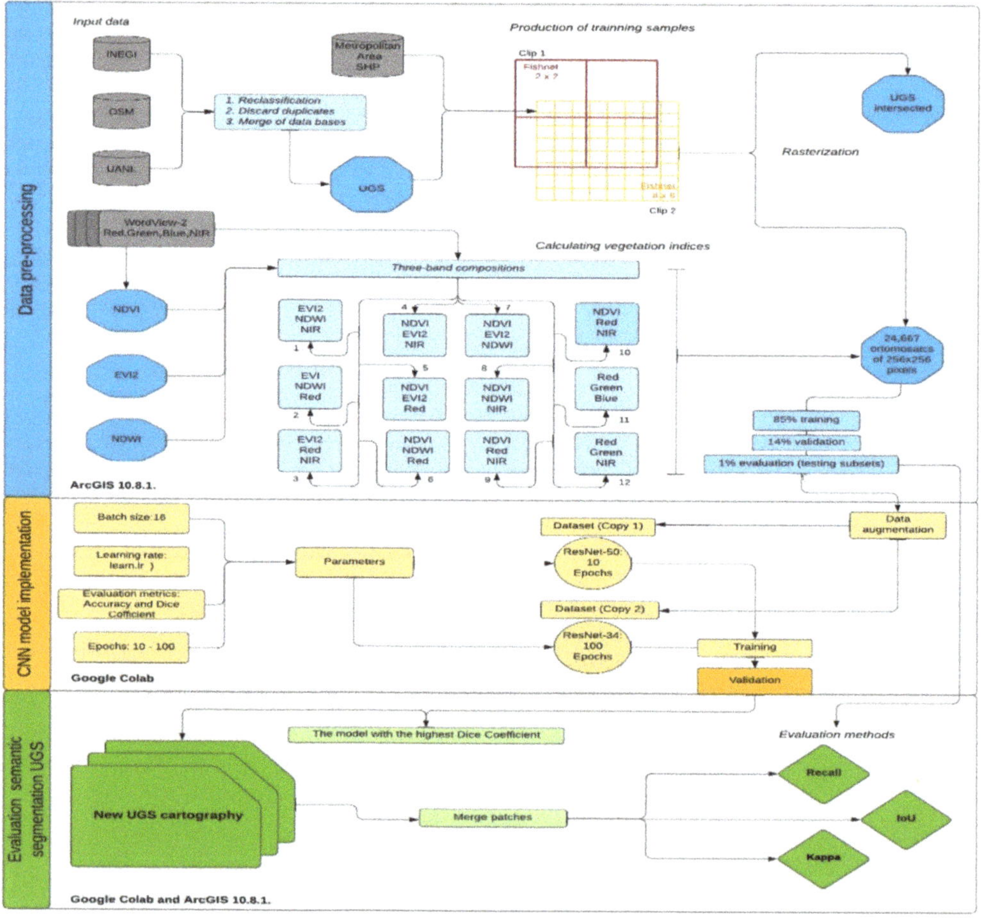

Figure 2. Summary of the current methodological process for semantic segmentation of UGSs using deep learning. Data preprocessing in blue, CNN model implementation in yellow, and evaluation of semantic segmentation of UGSs in green. Input data in gray, and all the processes appear in light blue, yellow, and green. Abbreviation: Urban Green Spaces (UGS), Convolutional Neural Networks (CNN).

2.2. Input Data

An orthomosaic of the MMA with a 0.5 m pixel resolution was used for the classification of UGSs (Figure 1). It was generated from nine WorldView-2 (WV2) satellite images obtained between June and October 2017. The spectral information it contained includes the red, green, blue (RGB) and near-infrared (NIR) bands in the ranges of 630–690 nm, 510–580 nm, 450–510 nm, and 770–895 nm, respectively.

2.3. Data Preprocessing

Information of the three original databases was reclassified based on the function and geometric structure of the UGSs to generate a common UGS database. The database includes polygons representing (1) median strips along streets and avenues, which are characterized by their elongated and narrow shapes; (2) residential gardens, which have pixels that correspond to vegetation managed by the municipality; (3) roundabouts, which have a round shape; (4) squares, which are spaces mostly used for recreation that maintain a symmetry and lack elements related to sports; (5) parks, which are embedded in residential areas, used for recreation and sports, and tend to be asymmetrical. Original classifications (some of them in Spanish) and their equivalent names after the reclassification process are shown in Table 1.

Table 1. Label reassignment on the reclassification process of the three databases.

Database	Original Fields		Reclassified Fields
Department of Geomatics (UANL)	NA [1]		Median strips
INEGI	Geográfico [1]	Tipo 1 [1]	
		Bordo [1]	Median strips
	Camellón [1]	Camellón [1]	Median strips
		Glorieta [1]	Roundabouts
		Área verde [1]	Parks
	Plaza [1]	NA [1]	Squares
	Instalación deportiva o recreative [1]	Parque [1]	Parks
		Jardín [1]	Residential gardens
OSM	Leisure		
	Playground		Parks
	Park		Parks
	Common		Parks

[1] Original data in Spanish.

Polygons that presented overlap were discarded with the employment of ArcMap "select by location" tool. The three reclassified databases were merged to produce the input shapefile for a rasterization process. The resulting product had a resolution of 0.5 m and was carried out for the generation of the final label raster. The pixel values determined the presence or absence of UGSs corresponding to median strips, roundabouts, parks, squares, and residential gardens. An additional sixth class named non-UGS was added to cover background pixels.

With the use of the green, red, and NIR bands, normalized difference vegetation index (NDVI) [59,60], two-band enhanced vegetation index (EVI2) [61,62], and normalized difference water index (NDWI) [63] were calculated using Equations (1)–(3), respectively.

$$\text{NDVI} = \frac{(NIR - Red)}{(NIR + Red)} \tag{1}$$

where *NIR* represents the near-infrared channel and *Red* is the red channel.

$$\text{EVI2} = 2.5 \frac{(NIR - Red)}{(NIR + 2.4 * Red + 1.0)} \tag{2}$$

where *NIR* represents the near-infrared channel and *Red* is the red channel.

$$\text{NDWI} = \frac{(Green - NIR)}{(Green + NIR)} \tag{3}$$

where *Green* represents the green channel and *NIR* is the near-infrared channel.

This study used 12 three-band compositions to determine their potential for the segmentation of UGSs. The bands used for the combinations were the produced indices (NDVI, NDWI, EVI2) and the spectral single bands NIR, red, green, and blue obtained from the original WV2 data (Figure 1).

As part of the process, 24,667 orthomosaics with dimensions of 256 × 256 pixels were produced from the label raster and each of the 12 three-band compositions was generated for the MMA. To obtain those orthomosaics, the original compositions and their respective label rasters were clipped using first a 2 × 2 mosaic fishnet (Clip1) with resulting orthomosaics that cover an area of 1336.64 km². Using the results of Clip1, a second fishnet of 8 × 8 mosaics (Clip2) was applied to each section to obtain 135 segments of 167.08 km², with 50 produced for the quadrant of the cardinal NE position, 43 for the NW, 9 for the SE, and 33 for the SW. Both fishnets developed for the generation of training samples from the MMA orthomosaic and UGS labels are shown in Figure 3A. Subsequent split raster process was performed in ArcMap, ArcGIS v10.8.1 software, for each of the quadrants previously generated, and over 24,000 orthomosaics were obtained for each of the three-band combinations as well as their equivalent ground truths. All the data had a spatial resolution of 0.5 m. The data were divided in a proportion of 85% for training and 14% for validation [64]. An additional 1% of the information was used for the evaluation of the model. The results were hosted on Google Drive's cloud storage service for later use through the Google Colab platform, which provided a Tesla P100 PCIe 16 GB GPU.

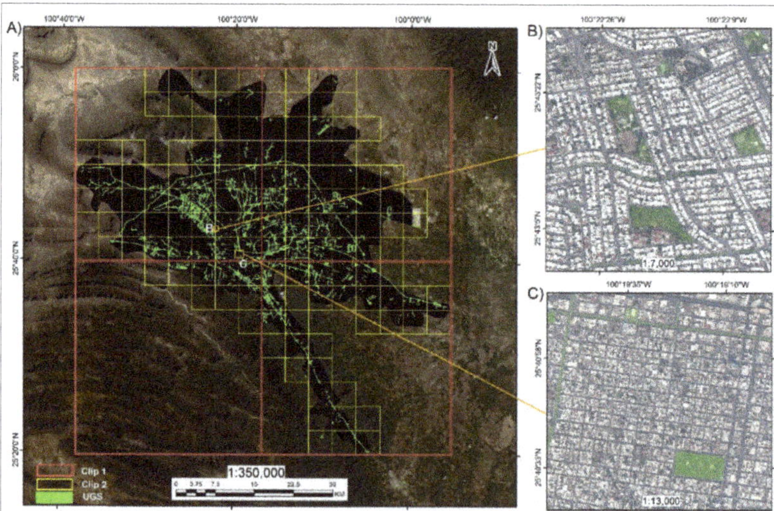

Figure 3. Production of training samples. (**A**) A 1:350,000 scale map with the result of data homogenization and rasterization of UGSs (light green polygons) with two raster extraction processes shown in red mosaics fishnet (Clip 1) and yellow fishnet (Clip 2). (**B**) Irregular geometry 1:7000 scale map of UGSs (light green polygons) extracted from Clip 2 and (**C**) Irregular geometry 1:13,000 scale map of UGSs (light green polygons) extracted from Clip 2.

2.4. CNN Model Implementation

Twenty-four semantic segmentation models were implemented via CNN, two for each band composition generated. ResNet-34 and ResNet50 encoders pre-trained by the

ImageNet dataset [65] were used in a dynamic U-Net architecture implemented in the fastai deep learning library [66]. This library works with the Python language and the PyTorch library [67] as a backend. Figure 4A shows the architecture for the model based on the pre-trained ResNet-34 encoder. Each model received 256 × 256 pixel images as input. ResNet network architectures integrate connection jumps (Figure 4B) that avoid the leak gradient problem present in other types of networks [68]. This helps to maintain performance and precision despite increases in the number of training layers [69]. At the point of the greatest compression in the FCN, a decoder was attached that follows the principle of the U-Net architecture to finally obtain an output equal in size to the input images.

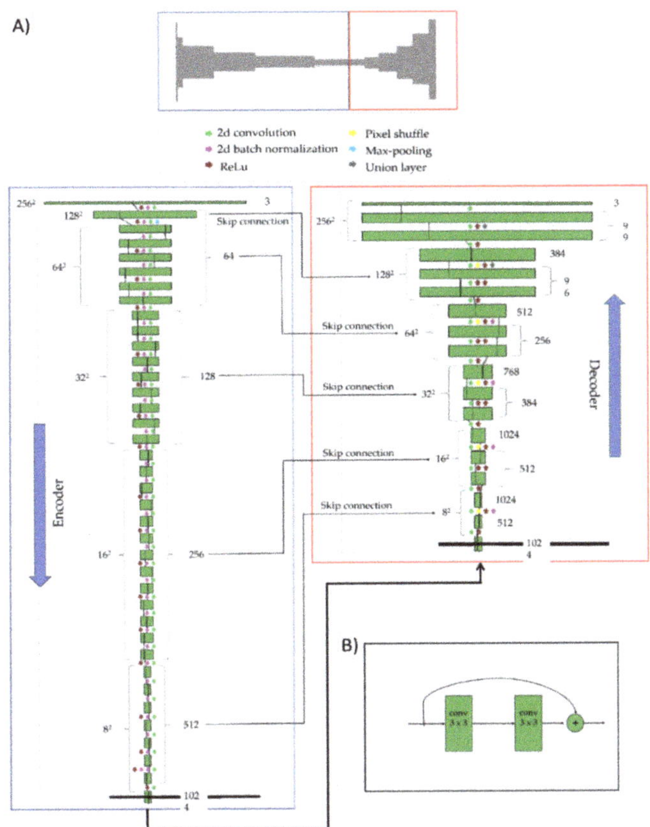

Figure 4. Dynamic U-Net model used for semantic segmentation. (**A**) The encoder consists of a ResNet-34 into which the orthomosaic and UGS labeling information is integrated. The numbers on the left of the graphs are the sizes of the input and output images, and the sizes of the activation maps. The numbers on the right are the channels/filters. (**B**) Building blocks used in the encoder section of the model.

Preliminary tests consisted of trial and error based on the limited literature related to UGS segmentation using deep learning models [38,70]. According to the capabilities of the system, a batch size of 16 samples was assigned. Data augmentation included (1) transformations with image turns at different angles, finding a 50% probability of being horizontal or vertical; (2) random symmetrical deformations with values of 0.1 magnitude; (3) random rotations with angles of 20°; (4) changes in the focus of images, up to 200%;

and (5) changes in light and contrast by a factor of 0.3. These techniques generated transformations for each epoch within the models and increased the size of the training samples by 60 times. Examples of image transformations can be observed in Figure 5.

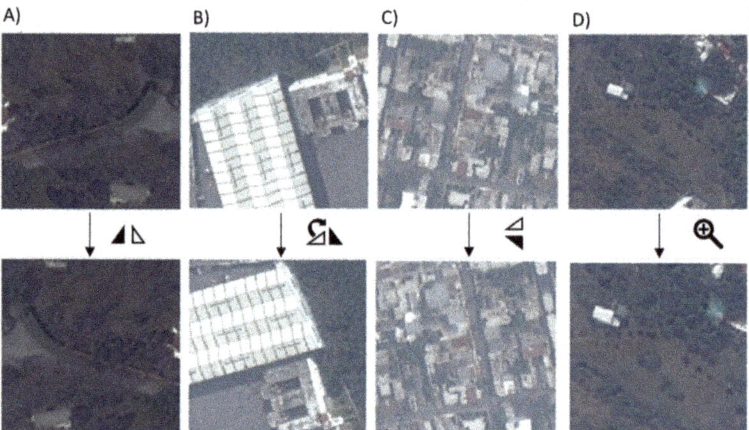

Figure 5. Example of images produced by data augmentation. (**A**) Horizontal flip; (**B**) random rotations; (**C**) vertical flip; (**D**) change in focus.

The accuracy score is an evaluation metric that quantifies the percentage of correctly classified pixels made by the predictions of the model [71]. It is calculated by Equation (4).

$$\text{Accuracy Score} = \frac{TP + TN}{TP + TN + FP + FN} \tag{4}$$

where TP represents the true positives, TN is the true negatives, FP is the false positives, and FN is the false negatives.

In semantic segmentation, the loss function metric is an algorithm used to evaluate the difference between training results and labeled data. To determine the most appropriate loss function metric for our data, we considered their spatial characteristics. As UGS represents a small portion of pixels at the metropolitan level, most cities are covered by built-up areas occupied by streets, buildings, and other impervious surfaces. This configuration causes an imbalance of classes that could produce errors and bias towards the background class that covers most of the area of interest. Semantic segmentation studies that used deep learning [72–74] have proved that the Dice coefficient or F1 score [75] is a loss function adequate for these kinds of problems. The Dice coefficient is calculated according to Equation (5).

$$\text{Dice Coefficient} = \frac{2|I_{GT} \cap O_{SEG}|}{|I_{GT}| + |O_{SEG}|} \tag{5}$$

where I_{GT} is the input ground truth and O_{SEG} is the output segmentation.

A Dice coefficient of 0 indicates that there is no overlap between the data, whereas a value of 1 means that the data has total overlap [76]. Because the input data consisted of three-band composed images, the Dice coefficient was computed for each class and then averaged via arithmetic mean through the fastai implementation [77]. The optimal learning rate for each model was defined using the *learn.lr_find()* method present in the fastai library [66]. This hyperparameter increases the learning rate from an exceptionally low value to the point where the loss gradient decreases [78]. Each ResNet34 model had 100 epochs to test the functionality of the implementations, this value was established according to the literature [79,80]. ResNet50 was implemented using 10 epochs due to reported Google Colab limitations on GPU, RAM, and session time availability. Mean, standard deviation, and confidence intervals of 95% were produced for the models to determine the statistically significant differences between the Dice coefficient calculated for each three-band combination.

2.5. Evaluation of Semantic Segmentation of UGSs

The model with the highest Dice coefficient was evaluated using the additional 1% testing subset taken from the original information. This subset, which was produced using the semantic segmentation, results in footprints used to obtain vector information corresponding to UGS polygons. Vectorization was implemented using Solaris library (https://solaris.readthedocs.io/, 26 November 2020) on Google Colab. The generated polygons were downloaded and then analyzed in ArcGIS to evaluate the model. The evaluation data contained 68 mosaic samples (Figure 6) with a total coverage of 32.08 hectares (ha). These samples permitted the evaluation of the effectiveness of the CNN with images different from those of the training and validation sets.

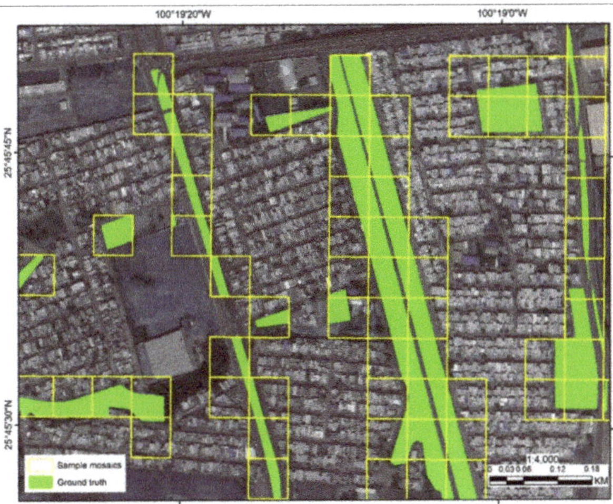

Figure 6. The ground truth coverage and the testing subset with 68 mosaics used to evaluate the CNN model.

As part of the evaluation, the intersection over union (IoU) was calculated. This metric computes the amount of overlap between the predicted polygons and the ground truth data [81] (Equation (6)).

$$\text{Intersection over Union} = \frac{|I_{GT} \cap O_{SEG}|}{|I_{GT} \cup O_{SEG}|} \tag{6}$$

where I_{GT} is the input ground truth and O_{SEG} is the output segmentation.

The recall analysis was obtained for the evaluation. This metric calculates the proportion of positives identified correctly as shown in Equation (7).

$$\text{Recall} = \frac{TP}{TP + FN} \tag{7}$$

where TP represents the true positives and FN is the false negatives.

The overall user and producer accuracies and the kappa coefficient were processed for the accuracy assessment of the evaluation data. This index of agreement is obtained through the computation of a confusion matrix with errors of omission and commission between classified maps and ground truth data; a kappa coefficient of 1 represents a perfect agreement, and a value of 0 indicates that the agreement is not as it was expected by chance [82,83]. For the calculation of the index, 2000 random points were generated by stratified random sampling and were labeled and verified against the reference data. The kappa coefficient (Equation (8)) is computed as follows:

$$\text{Kappa Coefficient} = \frac{N \sum_{i=1}^{r} x_{ii} - \sum_{i=1}^{r}(x_{i+} * x_{+i})}{N^2 - \sum_{i=1}^{r}(x_{i+} * x_{+i})} \tag{8}$$

where r represents the number of rows and columns in the error matrix, N is the total number of pixels, x_{ii} is the observation in row i and column i, x_{i+} is the marginal total of row i, and x_{+i} is the marginal total of column i.

3. Results

3.1. Data Preprocessing

The final dataset used as model input is shown in Table 2. It presents numbers of polygons, the area covered by the UGS classes within the MMA, and the proportions of area. The UGS pixel ratio was 3.003%. The data augmentation technique helped to improve the amount of supporting information for the training process by increasing the number of mosaics from 24,667 to 1,480,020 mosaics. With this increase, the CNN improved the learning process.

Table 2. Results of the homogenization of UGS databases.

UGS	Polygons	UGS Area (m^2)	Proportion (%)
Median strips	19,869	1,141,179	0.843
Residential gardens	1818	463,314.5	0.342
Roundabouts	61	810	0.001
Squares	58	14,076	0.010
Parks	2861	2,446,925	1.807
TOTAL	24,667	4,066,304.5	3.003

Parks represented the most prominent type of UGS with 1.8% coverage of the MMA (Table 2). Median strips represented 0.84% cover. Residential gardens represented 0.34% cover. The classes with the lowest coverage were squares and roundabouts with 0.01% and 0.001%, respectively.

3.2. Semantic Segmentation of UGSs

The highest Dice coefficient and accuracy results of the semantic segmentation for each of the 12 three-band compositions are presented in Table 3. NDVI–red–NIR composition achieved the best results using ResNet34 encoder with a Dice coefficient value of 0.5748 and an accuracy of 0.9503. Red–green–blue composition achieved the best results using ResNet50 with a Dice coefficient of 0.4378 and an accuracy of 0.9839. In contrast, EVI2–NDWI–NIR composition had the lowest values for both encoders. For the ResNet34 encoder model, the mean Dice coefficient was 0.49, the standard deviation was 0.09, and the statistical significance using 95% confidence intervals ranged from 0.42 to 0.55. For the ResNet50 encoder model, the mean Dice coefficient was 0.42, the standard deviation was 0.58, and the statistical significance using 95% confidence intervals ranged from 0.28 to 0.36.

Figure 7 illustrates the behavior of the training and validation process for the highest Dice coefficient for both encoders. As observed, the ResNet34 learning process extended to the 100 epochs (fluctuating between 0.45 and 0.63) and presented its peak at the 83rd epoch, reaching a Dice coefficient of 0.5748. In contrast, the ResNet50 learning process activity occurred during the first 10 epochs, reaching the best Dice coefficient of 0.4378 on the 4th epoch.

Table 3. Semantic segmentation model validation results for UGSs in VHR satellite images.

	ResNet34		ResNet50	
Band Compositions	**Dice Coefficient**	**Accuracy**	**Dice Coefficient**	**Accuracy**
EVI2–NDWI–NIR	0.1940	0.8853	0.2231	0.9065
EVI2–NDWI–Red	0.4961	0.9337	0.2543	0.9147
EVI2–Red–NIR	0.5113	0.942	0.3199	0.9145
NDVI–EVI2–NIR	0.5307	0.9437	0.2698	0.9074
NDVI–EVI2–Red	0.5021	0.9452	0.3356	0.9227
NDVI–NDWI–Red	0.5248	0.9433	0.3187	0.9115
NDVI–EV2–NDWI	0.4617	0.9347	0.3548	0.9249
NDVI–NDWI–NIR	0.4886	0.9377	0.2763	0.9369
NDVI–Red–NIR	0.5748	0.9503	0.3149	0.9004
NDWI–Red–NIR	0.5702	0.9505	0.3610	0.9200
Red–Green–Blue	0.4638	0.9792	0.4378	0.9839
Green–Red–NIR	0.5193	0.9547	0.3663	0.9322

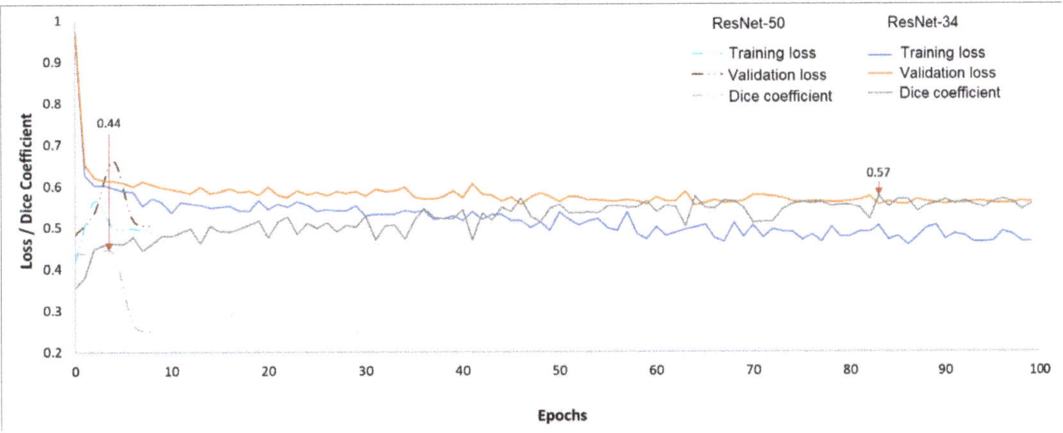

Figure 7. Plot of training loss, validation loss, and Dice coefficient for both encoders.

The best segmentation process is represented by the lowest loss value. An example of this is shown in Figure 8A–D, where NDVI–red–NIR composition using ResNet34 reflects how the learning process increases from a loss of 1.14 to 0.77. This behavior is also observed for the RGB combination using ResNet50 where the loss was from 1.47 to 0.84 (Figure 8E–H).

IoU metric was 0.75 for the evaluation of the NDVI–red–NIR composition. This was calculated using the polygons presented in Figure 9A. The recall analysis revealed that the ground truth data had an overlap of 96.07% with the predicted data and the proportion of the overlapping polygons corresponding to the predicted data was 80.04% (Figure 9B).

Results of the confusion matrix and kappa coefficient produced for the evaluation dataset are shown in Table 4. Both the ground truth and the predicted data contained polygons corresponding to parks and median strips classes. The kappa coefficient was 0.94, and the overall accuracy calculated was 0.97. The user accuracy was 1 for the parks and 0.96 for median strips, and the producer accuracy was 0.92 for the parks and 1 for median strips.

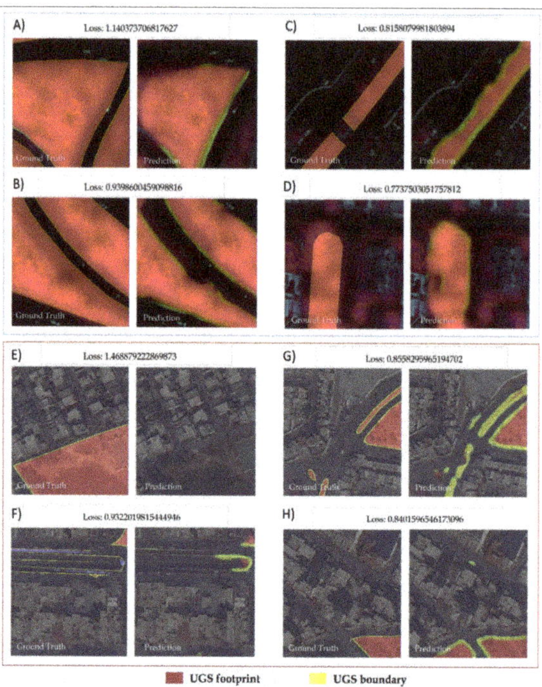

Figure 8. Progression of the segmentation process of UGSs for both encoders. (**A–D**) Samples obtained using an NDVI–red–NIR composition from ResNet34 encoder. (**E–H**) Samples obtained using a red–green–blue composition from ResNet50 encoder. The information is presented at the same scale; each square surface is 1.63 hectares (ha).

B)		Recall	
	T (ha)	Ground truth vs predicted (ha)	Ground truth vs predicted (%)
Ground truth	9.73	9.35	96.07
Predicted	11.68	9.35	80.04

Figure 9. Model assessment. (**A**) The overlap coverage between the polygons produced from the CNN model and the ground truth data. (**B**) Recall analysis. T is the total coverage. Ground truth vs predicted columns show the area in ha and proportion of positives identified correctly.

Table 4. Confusion matrix and kappa coefficient.

	Parks	Median Strips	Total	User Accuracy	Kappa Coefficient
Parks	684	0	684	1	0
Median strips	58	1258	1316	0.96	0
Total	742	1258	2000	0	0
Producer accuracy	0.92	1	0	0.97	0
Kappa coefficient	0	0	0	0	0.94

4. Discussion

The two automated methods tested for the identification of individual UGS polygons at the metropolitan level generated new databases that provide useful information, including geometry, condition, and spatial attributes, for the decision-making process regarding these important open public spaces. Updated databases improve national inventories with detailed geospatial and geometric information of UGSs, which is important for assessing their distribution and management. Updated and accurate UGS information, however, is difficult to acquire or access, especially in developing countries with no access to VHR imagery. Latin American countries lack this kind of information, where the UGS inventories are typically based on photo-interpretation techniques and depend on the user experience.

The information produced with this methodology can be used in conjunction with demographic information to analyze the accessibility and connectivity of UGSs at the metropolitan level. While this is a portion of the information needed for the quantification of the achievement of the SDG 11.7, the generation of a similar database considering non-green public spaces should also be contemplated to cover the analysis of both elements.

The methodology of segmentation of UGS polygons at the city level proposed in this work will allow effectively updating this information for urban spaces in Mexico. The method includes the typical UGS classes, such as median strips, roundabouts, parks, squares, and residential gardens present in every metropolitan area. OSM and INEGI open access databases used in this research are available for all of Mexico. This information was complemented by information produced by the local university through a metropolitan project funded by the state government, proving the importance of integrating multilevel governance (or institutions) to enhance and update geospatial data such as UGS inventories.

Methods to increase data representation are a necessity when VHR imagery is limited. In this work, data augmentation techniques using simple strategies showed their effectiveness by providing over 1 million additional orthomosaics. The variations that occurred in each of these image transformations helped to reach a more complete training set representing the complexity occurring in the study area due to temporal or environmental disparities.

The prospected combinations produced by four-band VHR imagery and their implementation using two encoders allowed the assessment of 24 segmentation models. This kind of modeling is only possible with a high computation capability. When that is not available, other options such as deep learning processing cloud services (e.g., Google Colab, Amazon AWS, Microsoft Azure) can be implemented.

According to the semantic segmentation model validation results and its statistical analysis, there is a significant difference (95%) between the dice coefficient of the different band combinations in both models. The best three-band combination for the semantic segmentation of UGSs is NDVI–red–NIR when using ResNet34 encoder and red–green–blue when using ResNet50 encoder. A future analysis regarding the learning process could help to identify the learning patterns and the influence of each band within the models by using interpretability, representation learning, and visualization methods [84].

A large difference between the validation accuracy and the dice coefficient was observed in Table 3. This variance is associated with the data imbalance caused by background non-UGS pixels. As the non-UGS class has the highest number of pixels in the analyzed orthomosaics, the accuracy of the model is high as it is quantifying a high percentage of correctly classified pixels for the entire area. The dice coefficient is a more reliable parameter in semantic segmentation processes with data imbalances because it reflects a metric based only on the segmented classes.

Semantic segmentation studies using similar approaches to map urban tree coverage, buildings, and roads [85–87] reported dice coefficients of 0.94, 0.84, and 0.87, respectively. The result in this study is lower (0.57), which may be related to the complexity of mapping UGS polygons. The referenced studies focus on the segmentation of classes that represent the city coverage; however, this research seeks to segment the pixel class and also several types of geometry. Additionally, UGS polygons are composed of a mix of pixels representing not only vegetation but also other kinds of infrastructure, such as sidewalks and playgrounds, which decrease the certainty for the segmentation process.

The kappa coefficient produced in the accuracy assessment of the evaluation data indicated a strong agreement between the predicted polygons and the ground truth data. With a value of 0.94, the kappa coefficient was similar to high accuracy results obtained in recent studies related to UGS mapping methods [88,89]. This indicates that the methodology used in this study is accurate for extracting and updating geometrical UGS databases at the metropolitan level.

5. Conclusions

This study evaluated two deep learning model techniques for semantic segmentation of UGS polygons with the use of different CNN encoders on the U-Net architecture to improve the methodology of UGS cartography. The models have the capability to detect patterns for all types of UGSs reported in Mexico, even with a high variation in shape or size, and to segment hundreds of thousands of polygons that represented 3% of the total MMA.

Results demonstrate that this methodology is an accurate digital tool for extracting and updating geometrical UGS databases at the metropolitan level (Dice coefficient of 0.57, recall of 0.8, IoU of 0.75, and kappa coefficient of 0.94). The implementation of these models could update UGS inventories necessary to assess urban management as cities grow or change. This methodology produces UGS geospatial data that are essential for quantifying the accomplishment of the SDG 11.7 regarding green spaces. This information in combination with demographic data could be used to elaborate UGS accessibility maps necessary to assess UGS accessibility. This new cartography may improve urban management for the conservation of natural resources and the environmental services they provide, as well as making their maps more accessible to urban residents and decision-makers.

Author Contributions: Conceptualization, F.D.Y.; data curation, R.E.H.; formal analysis, R.E.H. and F.D.Y.; funding acquisition, F.D.Y.; investigation, R.E.H. and F.D.Y.; methodology, R.H.G. and F.D.Y.; project administration, F.D.Y.; software, R.E.H.; supervision, F.D.Y., R.A.C.G., D.F.L.-G., V.H.G.C., A.L.F.F., and H.d.L.G.; validation, F.D.Y. and D.F.L.-G.; visualization, R.H.G. and F.D.Y.; writing—original draft, R.H.G. and F.D.Y.; writing—review and editing, D.F.L.-G., V.H.G.C., A.L.F.F., H.d.L.G., R.A.C.G. and A.V.M. All authors have read and agreed to the published version of the manuscript.

Funding: This research was funded by the National Council of Science and Technology, grant number 559018 and grant for international research (Foreign Mobility 2019-1).

Acknowledgments: The authors extend their sincere thanks to Gustau Camps-Valls, researcher of the Image Processing Lab of the University of Valencia, who guided Roberto Huerta in neural network subjects through a stay in the Image Processing Laboratory. Map data are copyrighted by OpenStreetMap contributors and available from https://www.openstreetmap.org, (accessed on 26 November 2020). The authors thank William D. Eldridge for his support and comments.

Conflicts of Interest: The authors declare no conflict of interest.

References

1. Degerickx, J.; Hermy, M.; Somers, B. Mapping functional urban green types using high resolution remote sensing data. *Sustainability* **2020**, 2144. [CrossRef]
2. Zhang, Y.; Van den Berg, A.E.; Van Dijk, T.; Weitkamp, G. Quality over quantity: Contribution of urban green space to neighborhood satisfaction. *Int. J. Environ. Res. Public Health* **2017**, 535. [CrossRef]
3. Gómez-Baggethun, E.; Barton, D.N. Classifying and valuing ecosystem services for urban planning. *Ecol. Econ.* **2013**. [CrossRef]
4. Nastran, M.; Kobal, M.; Eler, K. Urban heat islands in relation to green land use in European cities. *Urban. For. Urban. Green.* **2019**. [CrossRef]
5. Bao, T.; Li, X.; Zhang, J.; Zhang, Y.; Tian, S. Assessing the Distribution of Urban Green Spaces and its Anisotropic Cooling Distance on Urban Heat Island Pattern in Baotou, China. *ISPRS Int. J. Geo Inf.* **2016**, 12. [CrossRef]
6. Matos, P.; Vieira, J.; Rocha, B.; Branquinho, C.; Pinho, P. Modeling the provision of air-quality regulation ecosystem service provided by urban green spaces using lichens as ecological indicators. *Sci. Total Environ.* **2019**. [CrossRef] [PubMed]
7. Gupta, K.; Roy, A.; Luthra, K.; Maithani, S. Mahavir GIS based analysis for assessing the accessibility at hierarchical levels of urban green spaces. *Urban. For. Urban. Green.* **2016**, *18*. [CrossRef]
8. Wu, J.; He, Q.; Chen, Y.; Lin, J.; Wang, S. Dismantling the fence for social justice? Evidence based on the inequity of urban green space accessibility in the central urban area of Beijing. *Environ. Plan. B Urban. Anal. City Sci.* **2018**. [CrossRef]
9. Ma, F. Spatial equity analysis of urban green space based on spatial design network analysis (sDNA): A case study of central Jinan, China. *Sustain. Cities Soc.* **2020**. [CrossRef]
10. Martins, B.; Nazaré Pereira, A. Index for evaluation of public parks and gardens proximity based on the mobility network: A case study of Braga, Braganza and Viana do Castelo (Portugal) and Lugo and Pontevedra (Spain). *Urban. For. Urban. Green.* **2018**, *34*, 134–140. [CrossRef]
11. Green, O.O.; Garmestani, A.S.; Albro, S.; Ban, N.C.; Berland, A.; Burkman, C.E.; Gardiner, M.M.; Gunderson, L.; Hopton, M.E.; Schoon, M.L.; et al. Adaptive governance to promote ecosystem services in urban green spaces. *Urban. Ecosyst.* **2016**. [CrossRef]
12. Sikorska, D.; Łaszkiewicz, E.; Krauze, K.; Sikorski, P. The role of informal green spaces in reducing inequalities in urban green space availability to children and seniors. *Environ. Sci. Policy* **2020**. [CrossRef]
13. Wolch, J.R.; Byrne, J.; Newell, J.P. Urban green space, public health, and environmental justice: The challenge of making cities «just green enough». *Landsc. Urban. Plan.* **2014**, *125*, 234–244. [CrossRef]
14. Recio, A.; Linares, C.; Banegas, J.R.; Díaz, J. Impact of road traffic noise on cause-specific mortality in Madrid (Spain). *Sci. Total Environ.* **2017**. [CrossRef]
15. Kabisch, N. The Influence of Socio-economic and Socio-demographic Factors in the Association Between Urban Green Space and Health. In *Biodiversity and Health in the Face of Climate Change*; Springer: Cham, Switzerland, 2019.
16. Sperandelli, D.I.; Dupas, F.A.; Pons, N.A.D. Dynamics of urban sprawl, vacant land, and green spaces on the metropolitan fringe of São Paulo, Brazil. *J. Urban. Plan. Dev.* **2013**. [CrossRef]
17. Wu, H.; Liu, L.; Yu, Y.; Peng, Z. Evaluation and Planning of Urban Green Space Distribution Based on Mobile Phone Data and Two-Step Floating Catchment Area Method. *Sustainability* **2018**, *10*, 214. [CrossRef]
18. Mensah, C.A. Destruction of Urban Green Spaces: A Problem Beyond Urbanization in Kumasi City (Ghana). *Am. J. Environ. Prot.* **2014**. [CrossRef]
19. Zhou, D.; Zhao, S.; Liu, S.; Zhang, L. Spatiotemporal trends of terrestrial vegetation activity along the urban development intensity gradient in China's 32 major cities. *Sci. Total Environ.* **2014**, *488–489*, 136–145. [CrossRef]
20. Östberg, J.; Wiström, B.; Randrup, T.B. The state and use of municipal tree inventories in Swedish municipalities–Results from a national survey. *Urban. Ecosyst.* **2018**. [CrossRef]
21. Schneider, A.K.; Strohbach, M.W.; App, M.; Schröder, B. The «GartenApp»: Assessing and communicating the ecological potential of private gardens. *Sustainability* **2020**, 95. [CrossRef]
22. Taylor, L.; Hochuli, D.F. Defining greenspace: Multiple uses across multiple disciplines. *Landsc. Urban. Plan.* **2017**. [CrossRef]
23. Skokanová, H.; González, I.L.; Slach, T. Mapping Green Infrastructure Elements Based on Available Data, A Case Study of the Czech Republic. *J. Landsc. Ecol.* **2020**. [CrossRef]
24. Huerta-García, R.E.; Ramírez-Serrat, N.L.; Yépez-Rincón, F.D.; Lozano-García, D.F. Precision of remote sensors to estimate aerial biomass parameters: Portable LIDAR and optical sensors. *Rev. Chapingo Ser. Ciencias For. Ambient.* **2018**, *24*, 219–235. [CrossRef]
25. Yépez Rincón, F.D.; Lozano García, D.F. Mapeo del arbolado urbano con lidar aéreo. *Rev. Mex. Ciencias For.* **2018**. [CrossRef]
26. Vatseva, R.; Kopecka, M.; Otahel, J.; Rosina, K.; Kitev, A.; Genchev Rumiana Vatseva, S.; Genchev, S. Mapping urban green spaces based on remote sensing data: Case studies in Bulgaria and Slovakia. Proceedings of 6th International Conference on Cartography & GIS, Albena, Bulgary, 13–17 June 2016.
27. Atasoy, M. Monitoring the urban green spaces and landscape fragmentation using remote sensing: A case study in Osmaniye, Turkey. *Environ. Monit. Assess.* **2018**. [CrossRef]
28. Xue, J.; Su, B. Significant remote sensing vegetation indices: A review of developments and applications. *J. Sensors* **2017**, *17*. [CrossRef]
29. Shen, C.; Li, M.; Li, F.; Chen, J.; Lu, Y. Study on urban green space extraction from QUICKBIRD imagery based on decision tree. In Proceedings of the 2010 18th International Conference on Geoinformatics, Geoinformatics, Beijing, China, 18–20 June 2010.

30. Kopecká, M.; Szatmári, D.; Rosina, K. Analysis of Urban Green Spaces Based on Sentinel-2A: Case Studies from Slovakia. *Land* **2017**, *6*, 25. [CrossRef]

31. Sharifi, A.; Hosseingholizadeh, M. The Effect of Rapid Population Growth on Urban Expansion and Destruction of Green Space in Tehran from 1972 to 2017. *J. Indian Soc. Remote Sens.* **2019**. [CrossRef]

32. Gülçin, D.; Akpınar, A. Mapping Urban Green Spaces Based on an Object-Oriented Approach. *Bilge Int. J. Sci. Technol. Res.* **2018**, *2*, 71–81. [CrossRef]

33. Khryaschev, V.; Ivanovsky, L. Urban areas analysis using satellite image segmentation and deep neural network. In Proceedings of the E3S Web of Conferences, Divnomorskoe Village, Russia, 9–14 September 2019.

34. Chen, L.C.; Papandreou, G.; Kokkinos, I.; Murphy, K.; Yuille, A.L. Semantic image segmentation with deep convolutional nets and fully connected CRFs. In Proceedings of the 3rd International Conference on Learning Representations, ICLR 2015-Conference Track Proceedings, San Diego, CA, USA, 7–9 May 2015.

35. More, N.; Nikam, V.B.; Banerjee, B. Machine learning on high performance computing for urban greenspace change detection: Satellite image data fusion approach. *Int. J. Image Data Fusion* **2020**. [CrossRef]

36. Kranjcic, N.; Medak, D.; Zupan, R.; Rezo, M. Machine learning methods for classification of the green infrastructure in city areas. *Earth Environ. Sci.* **2019**, *362*, 012079. [CrossRef]

37. Aguilar, R.; Kuffer, M. Cloud computation using high-resolution images for improving the SDG indicator on open spaces. *Remote Sens.* **2020**, 1144. [CrossRef]

38. Xu, Z.; Zhou, Y.; Wang, S.; Wang, L.; Li, F.; Wang, S.; Wang, Z. A novel intelligent classification method for urban green space based on high-resolution remote sensing images. *Remote Sens.* **2020**, 3845. [CrossRef]

39. Shen, S.; Han, S.X.; Aberle, D.R.; Bui, A.A.; Hsu, W. An interpretable deep hierarchical semantic convolutional neural network for lung nodule malignancy classification. *Expert Syst. Appl.* **2019**. [CrossRef]

40. Wurm, M.; Stark, T.; Zhu, X.X.; Weigand, M.; Taubenböck, H. Semantic segmentation of slums in satellite images using transfer learning on fully convolutional neural networks. *ISPRS J. Photogramm. Remote Sens.* **2019**. [CrossRef]

41. Noh, H.; Hong, S.; Han, B. Learning deconvolution network for semantic segmentation. In Proceedings of the IEEE International Conference on Computer Vision, Santiago, Chile, 7–13 December 2015.

42. Ian, G.; Bengio, Y.; Courville, A. *Deep Learning*; MIT Press: Cambridge, MA, USA, 2016; ISBN 9780262035613.

43. Shelhamer, E.; Long, J.; Darrell, T. Fully Convolutional Networks for Semantic Segmentation. *IEEE Trans. Pattern Anal. Mach. Intell.* **2017**, *39*, 640–651. [CrossRef]

44. Ronneberger, O.; Fischer, P.; Brox, T. U-net: Convolutional networks for biomedical image segmentation. In Proceedings of the Medical Image Computing and Computer-Assisted Intervention—MICCAI 2015, Lecture Notes in Computer Science, Munich, Germany, 5–9 May 2015; pp. 234–241. [CrossRef]

45. Christ, P.F.; Elshaer, M.E.A.; Ettlinger, F.; Tatavarty, S.; Bickel, M.; Bilic, P.; Rempfler, M.; Hofmann, F.; Anastasi, M.D.; Sommer, W.H.; et al. Automatic Liver and Lesion Segmentation in CT Networks and 3D Conditional Random Fields. *Int. Conf. Med. Image Comput. Comput. Assist. Interv.* **2016**. [CrossRef]

46. Li, X.; Chen, H.; Qi, X.; Dou, Q.; Fu, C.W.; Heng, P.A. H-DenseUNet: Hybrid Densely Connected UNet for Liver and Tumor Segmentation from CT Volumes. *IEEE Trans. Med. Imaging* **2018**. [CrossRef]

47. Zhang, Z.; Liu, Q.; Wang, Y. Road Extraction by Deep Residual U-Net. *IEEE Geosci. Remote Sens. Lett.* **2018**. [CrossRef]

48. Xu, Y.; Wu, L.; Xie, Z.; Chen, Z. Building extraction in very high resolution remote sensing imagery using deep learning and guided filters. *Remote Sens.* **2018**, 1461. [CrossRef]

49. Rakhlin, A.; Davydow, A.; Nikolenko, S. Land cover classification from satellite imagery with U-net and lovász-softmax loss. In Proceedings of the IEEE Computer Society Conference on Computer Vision and Pattern Recognition Workshops, Salt Lake City, UT, USA, 18–22 June 2018.

50. Ahn, J.J.; Kim, Y.; Lucio, J.; Corley, E.A.; Bentley, M. Green spaces and heterogeneous social groups in the U.S. *Urban. For. Urban. Green.* **2020**. [CrossRef]

51. Klopp, J.M.; Petretta, D.L. The urban sustainable development goal: Indicators, complexity and the politics of measuring cities. *Cities* **2017**. [CrossRef]

52. Wendling, L.A.; Huovila, A.; zu Castell-Rüdenhausen, M.; Hukkalainen, M.; Airaksinen, M. Benchmarking nature-based solution and smart city assessment schemes against the sustainable development goal indicator framework. *Front. Environ. Sci.* **2018**. [CrossRef]

53. Rahman, K.M.A.; Zhang, D. Analyzing the level of accessibility of public urban green spaces to different socially vulnerable groups of people. *Sustainability* **2018**, *10*, 3917. [CrossRef]

54. Sedesol; Conapo; Inegi. *Delimitación de las Zonas Metropolitanas de México 2010*; Gobierno de la Ciudad de México: Mexico City, Mexico, 2012; ISBN 9786074271645.

55. Carmona, J.M.; Gupta, P.; Lozano-García, D.F.; Vanoye, A.Y.; Yépez, F.D.; Mendoza, A. Spatial and temporal distribution of PM2.5 pollution over Northeastern Mexico: Application of MERRA-2 reanalysis datasets. *Remote Sens.* **2020**, 2286. [CrossRef]

56. Gobierno de Monterrey. Reglamento De Protección Ambiental E Imagen Urbana De Monterrey. Available online: http://portal. monterrey.gob.mx/pdf/reglamentos/Reg_proteccion_ambiental.pdf (accessed on 26 November 2020).

57. Haklay, M.; Weber, P. OpenStreet map: User-generated street maps. *IEEE Pervasive Comput.* **2008**, *7*, 12–18. [CrossRef]

58. Yepez-Rincon, F.D.; Ferriño-Fierro, A.L.; Guerra-Cobián, V.H.; Limón-Rodríguez, B. Uso de sensores remotos y VANTs para la gestion de areas verdes urbanas. 13° COBRAC-Congresso de Cadastro Multifinalitário e Gestão TerritorialAt: Florianápolis/Santa Catarina, Brazil, Florianápolis/Santa Catarina, Brazil, 21–24 October 2018.

59. Bannari, A.; Morin, D.; Bonn, F.; Huete, A.R. A review of vegetation indices. *Remote Sens. Rev.* **1995**. [CrossRef]

60. Gillespie, T.W.; Ostermann-Kelm, S.; Dong, C.; Willis, K.S.; Okin, G.S.; MacDonald, G.M. Monitoring changes of NDVI in protected areas of southern California. *Ecol. Indic.* **2018**. [CrossRef]

61. Huete, A.R.; Liu, H.Q.; Batchily, K.; Van Leeuwen, W. A comparison of vegetation indices over a global set of TM images for EOS-MODIS. *Remote Sens. Environ.* **1997**. [CrossRef]

62. Nepita-Villanueva, M.R.; Berlanga-Robles, C.A.; Ruiz-Luna, A.; Morales Barcenas, J.H. Spatio-temporal mangrove canopy variation (2001–2016) assessed using the MODIS enhanced vegetation index (EVI). *J. Coast. Conserv.* **2019**. [CrossRef]

63. Zylshal, S.S.; Yulianto, F.; Nugroho, J.T.; Sofan, P. A support vector machine object based image analysis approach on urban green space extraction using Pleiades-1A imagery. *Model. Earth Syst. Environ.* **2016**, *2*, 54. [CrossRef]

64. Benjelloun, M.; El Adoui, M.; Larhmam, M.A.; Mahmoudi, S.A. Automated Breast Tumor Segmentation in DCE-MRI Using Deep Learning. In Proceedings of the 4th International Conference on Cloud Computing Technologies and Applications, Cloudtech 2018, Brussels, Belgium, 26–28 November 2018.

65. He, K.; Zhang, X.; Ren, S.; Sun, J. Deep residual learning for image recognition. In Proceedings of the IEEE Computer Society Conference on Computer Vision and Pattern Recognition, Las Vegas, NV, USA, 26 June–1 July 2016.

66. Howard, J.; Gugger, S. Fastai: A layered api for deep learning. *Information* **2020**, *108*. [CrossRef]

67. Paszke, A.; Gross, S.; Massa, F.; Lerer, A.; Bradbury, J.; Chanan, G.; Killeen, T.; Lin, Z.; Gimelshein, N.; Antiga, L.; et al. PyTorch: An Imperative Style, High-Performance Deep Learning Library. *arXiv* **2019**, arXiv:1912.01703, preprint.

68. Balduzzi, D.; Frean, M.; Leary, L.; Lewis, J.P.; Ma, K.W.D.; McWilliams, B. The shattered gradients problem: If resnets are the answer, then what is the question? In Proceedings of the 34th International Conference on Machine Learning, ICML 2017, Sydney, Australia, 6–11 August 2017.

69. Wu, S.; Zhong, S.; Liu, Y. Deep residual learning for image steganalysis. *Multimed. Tools Appl.* **2017**, 1–17. [CrossRef]

70. Moreno-Armendáriz, M.A.; Calvo, H.; Duchanoy, C.A.; López-Juárez, A.P.; Vargas-Monroy, I.A.; Suarez-Castañon, M.S. Deep green diagnostics: Urban green space analysis using deep learning and drone images. *Sensors* **2019**, *5287*. [CrossRef]

71. Saraiva, M.; Protas, É.; Salgado, M.; Souza, C. Automatic mapping of center pivot irrigation systems from satellite images using deep learning. *Remote Sens.* **2020**, *558*. [CrossRef]

72. Huang, Q.; Sun, J.; Ding, H.; Wang, X.; Wang, G. Robust liver vessel extraction using 3D U-Net with variant dice loss function. *Comput. Biol. Med.* **2018**. [CrossRef]

73. White, A.E.; Dikow, R.B.; Baugh, M.; Jenkins, A.; Frandsen, P.B. Generating segmentation masks of herbarium specimens and a data set for training segmentation models using deep learning. *Appl. Plant. Sci.* **2020**. [CrossRef]

74. Liu, H.; Feng, J.; Feng, Z.; Lu, J.; Zhou, J. Left atrium segmentation in CT volumes with fully convolutional networks. In Proceedings of the 7th International Workshop, ML-CDS 2017, Québec City, QC, Canada, 10 September 2017.

75. Dice, L.R. Measures of the Amount of Ecologic Association Between Species. *Ecology* **1945**. [CrossRef]

76. Guindon, B.; Zhang, Y. Application of the Dice Coefficient to Accuracy Assessment of Object-Based Image Classification. *Can. J. Remote Sens.* **2017**. [CrossRef]

77. Opitz, J.; Burst, S. Macro F1 and Macro F1. *arXiv* **2019**, arXiv:1911.03347. Available online: https://arxiv.org/abs/1911.03347 (accessed on 26 November 2020).

78. Smith, S.L.; Kindermans, P.J.; Ying, C.; Le, Q.V. Don't decay the learning rate, increase the batch size. In Proceedings of the 6th International Conference on Learning Representations, ICLR 2018-Conference Track Proceedings, Vancouver, BC, Canada, 30 April–3 May 2018.

79. Zhang, X.; Xiao, Z.; Li, D.; Fan, M.; Zhao, L. Semantic Segmentation of Remote Sensing Images Using Multiscale Decoding Network. *IEEE Geosci. Remote Sens. Lett.* **2019**. [CrossRef]

80. Abraham, N.; Khan, N.M. A novel focal tversky loss function with improved attention u-net for lesion segmentation. In Proceedings of the International Symposium on Biomedical Imaging, Venice, Italy, 8–11 April 2019.

81. Stewart, E.L.; Wiesner-Hanks, T.; Kaczmar, N.; DeChant, C.; Wu, H.; Lipson, H.; Nelson, R.J.; Gore, M.A. Quantitative Phenotyping of Northern Leaf Blight in UAV Images Using Deep Learning. *Remote Sens.* **2019**, *2209*. [CrossRef]

82. Cohen, J. A Coefficient of Agreement for Nominal Scales. *Educ. Psychol. Meas.* **1960**, *20*, 37–46. [CrossRef]

83. Rwanga, S.S.; Ndambuki, J.M. Accuracy Assessment of Land Use/Land Cover Classification Using Remote Sensing and GIS. *Int. J. Geosci.* **2017**, *8*, 611–622. [CrossRef]

84. Janik, A.; Sankaran, K.; Ortiz, A. Interpreting Black-Box Semantic Segmentation Models in Remote Sensing Applications. In Proceedings of the EuroVis Workshop on Machine Learning Methods in Visualisation for Big Data, The Eurographics Association, Porto, Portugal, 3 June 2019.

85. Timilsina, S.; Aryal, J.; Kirkpatrick, J.B. Mapping urban tree cover changes using object-based convolution neural network (OB-CNN). *Remote Sens.* **2020**, *3017*. [CrossRef]

86. Khryashchev, V.; Larionov, R.; Ostrovskaya, A.; Semenov, A. Modification of U-Net neural network in the task of multichannel satellite images segmentation. In Proceedings of the 2019 IEEE East-West Design and Test Symposium, EWDTS 2019, Batumi, Georgia, 13–16 September 2019.

87. Molina, E.N.; Zhang, Z. Semantic segmentation of satellite images using a U-shaped fully connected network with dense residual blocks. In Proceedings of the 2019 IEEE International Conference on Multimedia and Expo Workshops, ICMEW, Shanghai, China, 8–12 July 2019.
88. Lahoti, S.; Kefi, M.; Lahoti, A.; Saito, O. Mapping Methodology of Public Urban Green Spaces Using GIS: An Example of Nagpur City, India. *Sustainability* **2019**, *11*, 2166. [CrossRef]
89. Sun, X.; Tan, X.; Chen, K.; Song, S.; Zhu, X.; Hou, D. Quantifying landscape-metrics impacts on urban green-spaces and water-bodies cooling effect: The study of Nanjing, China. *Urban. For. Urban. Green.* **2020**, *55*, 126838. [CrossRef]

remote sensing

Article

Susceptibility Analysis of the Mt. Umyeon Landslide Area Using a Physical Slope Model and Probabilistic Method

Sunmin Lee [1,2], Jungyoon Jang [3], Yunjee Kim [2], Namwook Cho [2] and Moung-Jin Lee [4,*]

1   Department of Geoinformatics, University of Seoul, 163 Seoulsiripdaero, Dongdaemun-gu, Seoul 02504, Korea; smilee@kei.re.kr
2   Environmental Assessment Group, Korea Environment Institute (KEI), 370 Sicheong-daero, Sejong-si 30147, Korea; yunjeekim@kei.re.kr (Y.K.); nwcho@kei.re.kr (N.C.)
3   Spatial Information Business Office Seoul Regional Headquarters, Korea Land and Geospatial Informatix Corporation, Seoul 06053, Korea; jyjang5904@lx.or.kr
4   Center for Environmental Data Strategy, Korea Environment Institute (KEI), 370 Sicheong-daero, Sejong-si 30147, Korea
*   Correspondence: leemj@kei.re.kr; Tel.: +82-44-415-7314

Received: 7 July 2020; Accepted: 13 August 2020; Published: 18 August 2020

Abstract: Every year, many countries carry out landslide susceptibility analyses to establish and manage countermeasures and reduce the damage caused by landslides. Because increases in the areas of landslides lead to new landslides, there is a growing need for landslide prediction to reduce such damage. Among the various methods for landslide susceptibility analysis, statistical methods require information about the landslide occurrence point. Meanwhile, analysis based on physical slope models can estimate stability by considering the slope characteristics, which can be applied based on information about the locations of landslides. Therefore, in this study, a probabilistic method based on a physical slope model was developed to analyze landslide susceptibility. To this end, an infinite slope model was used as the physical slope model, and Monte Carlo simulation was applied based on landslide inventory including landslide locations, elevation, slope gradient, specific catchment area (SCA), soil thickness, unit weight, cohesion, friction angle, hydraulic conductivity, and rainfall intensity; deterministic analysis was also performed for the comparison. The Mt. Umyeon area, a representative case for urban landslides in South Korea where large scale human damage occurred in 2011, was selected for a case study. The landslide prediction rate and receiver operating characteristic (ROC) curve were used to estimate the prediction accuracy so that we could compare our approach to the deterministic analysis. The landslide prediction rate of the deterministic analysis was 81.55%; in the case of the Monte Carlo simulation, when the failure probabilities were set to 1%, 5%, and 10%, the landslide prediction rates were 95.15%, 91.26%, and 90.29%, respectively, which were higher than the rate of the deterministic analysis. Finally, according to the area under the curve of the ROC curve, the prediction accuracy of the probabilistic model was 73.32%, likely due to the variability and uncertainty in the input variables.

Keywords: probabilistic method; Monte Carlo simulation; physical slope model; Mt. Umyeon landslides

---

1. Introduction

Landslides cause substantial economic and social losses, especially in urban areas where many people live. Landslides are destructive and represent the most frequent risk factors in mountainous areas; especially in urban areas where damage to forests and infrastructure, such as buildings and roads, can lead to soil erosion and thus land deformation. Therefore, analysis of mitigation risks and countermeasures for mitigation are essential steps in reducing natural disasters.

Geographic information system (GIS)-based landslide susceptibility analysis has been conducted to predict areas with high probabilities of landslides. Based on spatial data constructed from factors that influence landslide occurrence, such as topography, hydrology, forests, and geology, studies have been conducted to estimate landslide susceptibility [1–5]. Many traditional statistical methodologies, such as evident belief function, frequency ratio, analytical hierarchy process, and logistic regression have been applied [5–9]. Recently, machine learning methodologies have been applied to estimate the spatial uncertainty of landslides [1,4,10,11]. Typically, decision tree-based models [12], such as random forest and boosted tree models [13,14], have been applied. Support vector machines [15] and artificial neural networks [16,17] are other commonly applied machine learning methods for landslide susceptibility analysis. Various studies analyzed the relationship between landslide location data and the factors that cause landslides and calculate their effects on landslides. There is a disadvantage to this method because it is not possible to conduct landslide susceptibility analysis on target areas before landslides occur when there is no information on the locations of landslides. The landslide damage area in South Korea is expanding every year; thus, in addition to areas where landslides have already occurred, the frequency of new landslides is increasing [18]. Therefore, it is necessary to predict landslide susceptibilities and prepare countermeasures for areas without prior information.

Among the methods for evaluating landslide susceptibilities, physical-model based analysis estimates the stability by treating a slope as a specific physical model and inputting slope information [19,20]. This method enables susceptibility analysis regardless of the information about landslide occurrence location information, so it is possible to analyze the susceptibility of an area before landslides occur [21]. Therefore, this analysis method has the advantage of considering the occurrence mechanism and process of landslides and has been reported as one of the most effective techniques of landslide vulnerability and risk analysis [22]. In particular, among the physical slope models, the infinite slope model, which has a similar form of landslide fracture surface and is easier to analyze than other models, is most commonly used [23,24]. In addition, probabilistic techniques are used to effectively deal with the spatial variability in the geotechnical properties used as an input into physical slope models and inaccurate results due to complex geological conditions [25–33].

Landslide susceptibility analysis using physical models uses input data related to the topographical and geological characteristics of the slope [34]. In the process of obtaining the geotechnical characteristics, uncertainty occurs due to the spatial variability of the ground and complex geological conditions, which increases the possibility of obtaining incorrect analysis results. Therefore, probabilistic techniques such as Monte Carlo simulation have been used to quantify uncertainty [26,30,33]. In addition, studies are being conducted on various hydrogeological models capable of estimating pore water pressure due to rainfall infiltrating underground for the calculation of pore water pressure affecting stability by reducing shear strength of slope materials [35,36]

Thus, this study applied a physical model-based method to analyze landslide susceptibility before landslide occurrence. In this study, an infinite slope model was used as the physical slope model, and we used Monte Carlo (MC) simulation for the probabilistic analysis. Information on the actual occurrence of landslides in Mt. Umyeon was used to validate the accuracy of our techniques [37]. Figure 1 shows the detailed workflow used in this study. We used Mt. Umyeon as the study area because an urban landslide occurred here in July 2011. The Mt. Umyeon landslide is a representative example of serious human injury because it was located in the center of the metropolitan area of Seoul.

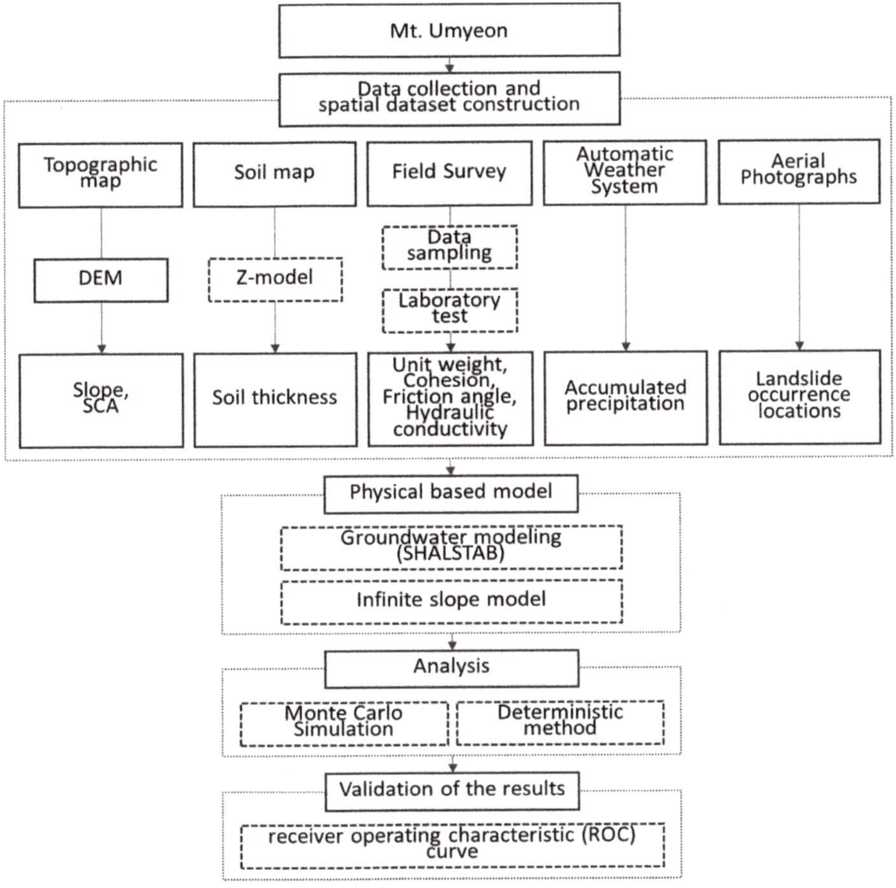

Figure 1. The workflow of this study.

2. Study Area

The study area, Mt. Umyeon, is located in Seoul, the capital of the Republic of Korea, within 126°59′–127°01′E, 37°27′–37°28′N (Figure 2). Mt. Umyeon is located at the center of a densely populated area, with highways to the east, rivers and parks to the south, and major cultural facilities around the Mt. Umyeon area. The study area, which was selected based on the type of watershed affected by rainfall, is approximately 22.18 km² (width: 5.075 km, length: 4.37 km). The maximum height of Mt. Umyeon is 293 m; the southern slope has a large slope and a valley, while the northern slope is gentle. Mt. Umyeon is a relatively low mountainous region of gneiss formed by retardation and weathering. The terrain is vulnerable to landslides because the gneiss in the bedrock is distributed with severe weathering and many faults. In addition, the dark veins are partially infiltrated, and overall, weathering is severe and the outcrop is poor [38].

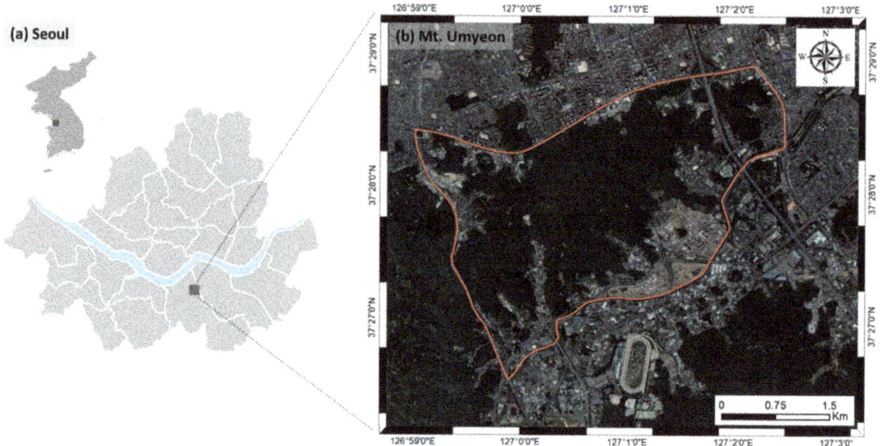

Figure 2. Study area: (**a**) Seoul; (**b**) Mt. Umyeon area.

In this region, a number of landslides occurred due to heavy rains, with a cumulative rainfall of 587.5 mm for 3 days from 26 to 28 July, 2011 [37]. About 150 large and small landslides occurred [38], and the area of debris flow was very wide compared to about 11 square kilometers selected as the radius of the study area (Figure 3). Landslides were presumed to result from heavy rains over a period of about 1 h following the weakening of ground due to previous heavy rains of 230.0–266.5 mm from about 15 h before the landslide [38]; the estimated time of the landslide was 09:00 on 27 July, 2011. A number of landslides in the form of debris flow have been reported and landslide occurrence locations were collected based on field investigation and visual analysis of aerial photographs before and after the landslides [38] by points [3,5,13].

Figure 3. Landslide status on the northern side of Mt. Umyeon in 2011 [39,40].

3. Spatial Datasets

Table 1 shows the spatial datasets used in this study. To build the input dataset, all input data were constructed in a raster format with a 5 × 5 m grid structure using the ArcGIS 10.3 software. For the analysis of landslide susceptibilities, relevant factors were selected through literature review of previous studies [41–46]. First, a 1: 5000 topographic map was obtained, from which we collated data such as elevation, slope, and specific catchment area. A digital elevation model was constructed by extracting a contour vector layer, including elevation attributes, from a digital topographic map

(Figure 4a). Finally, the slope and specific catchment area (SCA) were constructed based on the digital elevation model (Figure 4b,c).

Table 1. Landslide-related factors used to construct spatial database.

Data Source	Factors	Data Type	Scale
Aerial photograph [a]	Landslide location	Point	-
Topographical map [b]	Elevation [m] Slope gradient [°] Specific catchment area (SCA)	GRID	1:5000
Soil map [c]	Soil thickness [m]	Polygon	1:25,000
Field Investigation [d]	Unit weight [kN/m^3] Cohesion [kPa] Friction angle [degree] Hydraulic conductivity [cm/s]	Point	-
Precipitation [e]	Rainfall intensity	Point	-

[a] Aerial photograph before and after Mt.Umyeon landslides from Kakaomap [47]. [b] Topographical factors were extracted from digital topographic map by National Geographic Information Institute. [c] The detailed soil map produced by Rural Development Administration. [d] Field investigation data produced by Korean Geotechnical Society [38]. [e] The 16-h accumulated precipitation of from seven Automatic Weather System (AWS) observatories by Korea Meteorological Administration (KMA).

Second, the z-model was used to build the depth data of the fracture surface. The z-model is a submarine model that reflects the topographic characteristics of the slope by calculating its depth with respect to the elevation [48]. According to the results of ground drilling and seismic surveys by the Korean Society of Civil Engineers [37], the distribution of the soil thickness in the Mt. Umyeon area is 1.18–10.13 m. Therefore, the minimum and maximum depths of the soil in the z-model were set to 1 and 10 m, respectively, and soil thickness was calculated (Figure 4d) [37].

Third, unit weight, cohesion, friction angle, and hydraulic conductivity were obtained through field surveys with indoor experiments. Direct shear tests were conducted on four sampling points (SPs) on the obtained samples of cohesion and friction angle, and borehole shear tests and in-situ permeability tests were conducted at eight drilling points (DPs) in the study area (Figure 5) [38]. Table 2 summarizes the results obtained from this process. In order to interpolate and analyze geological characteristics of the entire study area based on the obtained data [32,41], kriging spatial interpolation analysis from ArcGIS, a spatial processing interpolation technique that reflects the correlation between the distance from the surrounding value and the value located around it, was performed to construct data in raster format (Figure 6).

Table 2. Geotechnical properties of unit weight, cohesion, friction angle, and hydraulic conductivity.

Name of Sampling Point	Unit Weight [kN/m^3]	Cohesion [kPa]	Friction Angle [degree]	Hydraulic Conductivity [cm/s]
SP-1	13.770	-	-	-
SP-2	13.530	-	-	-
SP-3	11.405	-	-	-
SP-4	12.405	-	-	-
DP-1	-	7.45	22.34	4.67×10^{-4}
DP-2	-	6.86	25.11	8.08×10^{-4}
DP-3	-	11.89	27.01	7.92×10^{-4}
DP-4	-	8.36	24.78	8.08×10^{-4}
DP-5	-	9.73	25.75	3.54×10^{-4}
DP-6	-	7.51	24.70	4.45×10^{-4}
DP-7	-	11.24	27.14	2.08×10^{-4}
DP-8	-	20.06	29.42	1.80×10^{-4}

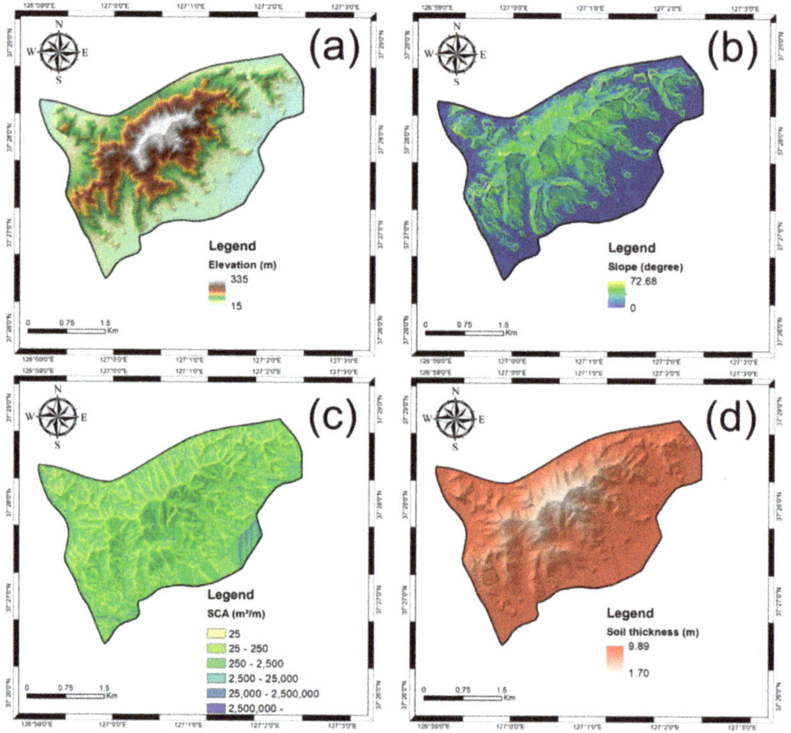

Figure 4. Map showing the input parameters constructed using topographic maps: (**a**) elevation; (**b**) slope; (**c**) specific catchment area (SCA); (**d**) soil thickness.

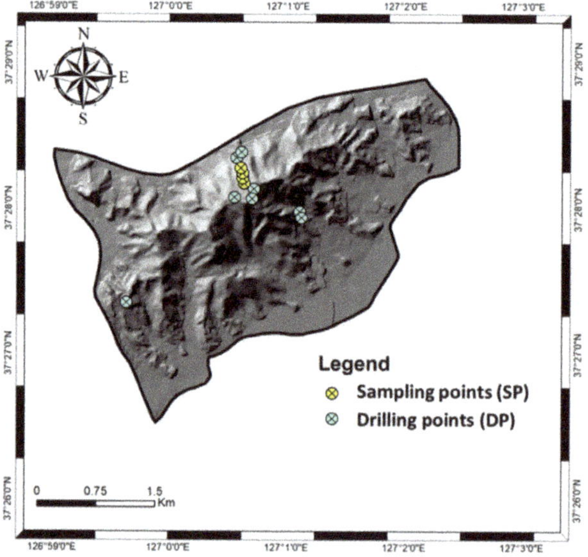

Figure 5. Location map of sampling point for acquisition of geotechnical properties.

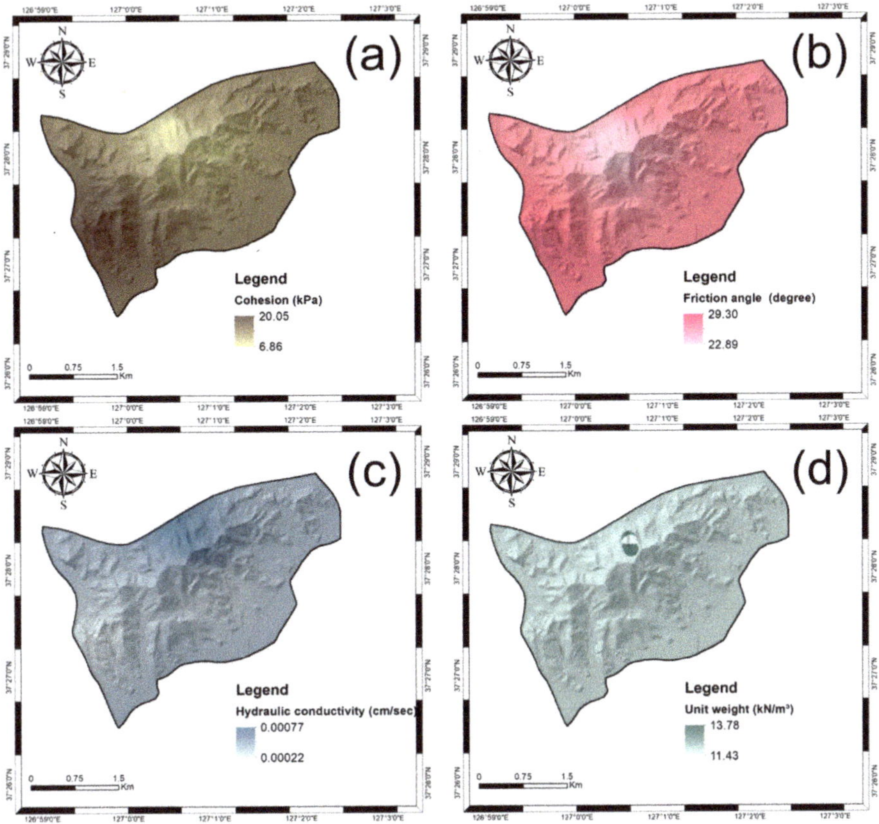

Figure 6. Map showing input parameters constructed using kriging interpolation based on the geotechnical properties of the sampling location (**a**) cohesion; (**b**) friction angle; (**c**) hydraulic conductivity; (**d**) unit weight.

In sequence, automatic weather system (AWS) data were used to construct rainfall intensity data. The Korea Meteorological Administration (KMA) operates over 510 AWS sites to monitor the atmospheric conditions near the ground in real time. To construct rainfall data in the study area, rainfall data was measured at 1-h intervals at seven AWS stations near Mt. Umyeon (Table 3). The rainfall intensity over the 16 h from 19:00 26 July to 09:00 27 July, 2011 was obtained to calculate the rainfall intensity at each AWS. Kriging interpolation was also applied to construct rainfall intensity data for the entire study area (Figure 7).

Table 3. Automatic weather system (AWS) observatory information.

AWS Observatory Name	Number of Stations	Latitude	Longitude	Height above Sea Level (m)
Gwanak(ra)	116	37.44526	126.96402	625
Gangnam	400	37.5134	127.04671	59
Seocho	401	37.48462	127.02601	33
Yongsan	415	37.52038	126.97611	31.73
Namhyeon	425	37.46336	126.9855	88
Gwanak	509	37.45284	126.95015	142
Gwacheon	590	37.44028	127.00249	47

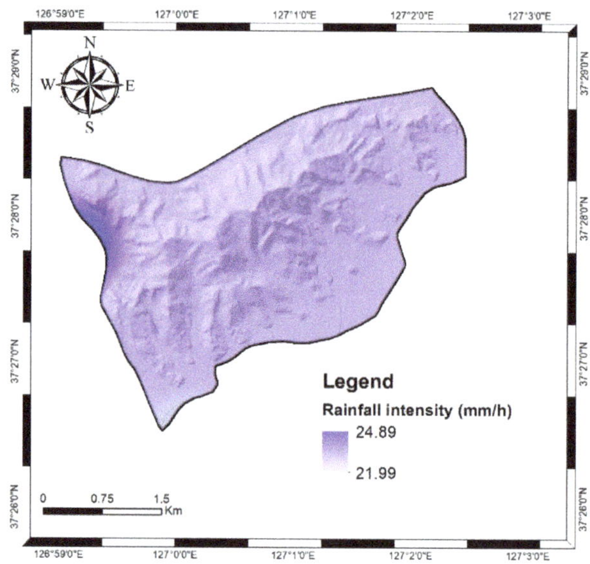

Figure 7. Map showing rainfall intensity using kriging interpolation based on automatic weather system observatory locations.

Finally, a 2011 landslide occurrence map was prepared to validate the analysis. A total of 103 landslide locations were extracted by superimposing a 1: 5000 digital topography onto an aerial photograph with a spatial resolution of 50 cm before and after the landslide (Figure 8).

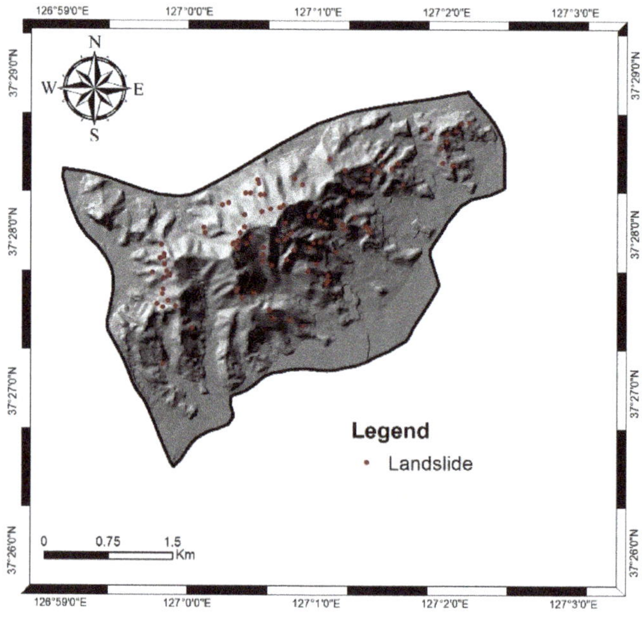

Figure 8. Landslide locations.

4. Methodology

4.1. Physically Based Model

We used a physical slope model, an infinite slope, in which the depth of the ground is shorter than the length of the landslide. The infinite slope model is the most suitable model for GIS-based landslide analysis and is widely used to analyze landslides caused by rainfall [49–52]. The stability of the infinite slope model is expressed by the factor of safety (FS), which is the ratio of shear stress to shear strength. The shear stress is the stress that acts on the fracture surface, taking into account the weight and influence of the slope material, whereas the shear strength is based on the action of the pore water pressure on the vertical stress acting perpendicularly to the fracture surface. Therefore, an FS value of less than 1 indicates a destructive state; and is expressed as [53]:

$$\mathrm{FS} = \frac{c + (\gamma D - \gamma_w z_w)\, cos^2 \alpha \tan \varnothing}{\gamma D \sin \alpha \cos \alpha} \tag{1}$$

where c is the cohesion [kN/m^3], γ is the unit weight of the slope material [kN/m^2], D is the depth of the slope [m], γ_w is the unit weight of water [kN/m^2], z_w is the groundwater level [m], α is a gradient of the slope [degree], and φ is a frictional angle [degree]. The SHALlow landsliding STABility model (SHALSTAB) was used, which is based on the distributed hydrological model [36] and wet index of the infinite slope stability model, to calculate the groundwater level (z_w). SHALSTAB is a hydraulic model that considers only the ground flow while ignoring the outflow of the surface in the hydraulic model proposed by [54], which considers the flow and surface runoff in shallow ground. It is expressed as follows:

$$\frac{q \times a}{T \times \sin \alpha \times b} = \frac{z_w}{D} = w \tag{2}$$

where q is rainfall [m/day], a is the watershed area [m^2], b is the width of the contour line [m], and T is transmissivity [m^2/day]. Furthermore, w is the wetness index, which is the ratio of the groundwater level to the depth of the slope (z_w/D), the relative depth of the actual groundwater level with respect to the slope. The wetness index w is between 0 and 1; its maximum value is 1 because the groundwater level does not exceed the depth of the slope. The groundwater level is expressed as:

$$w = \mathrm{Min} \left(\frac{q \times a}{T \times \sin \alpha \times b}, 1 \right) \tag{3}$$

4.2. Monte Carlo Simulation

In order to calculate the safety factor through a mathematical model, input data such as shear strength (cohesion force and friction angle) of the ground or groundwater level are required. However, these data are usually obtained through limited field surveys or indoor experiments, which is absolutely insufficient compared to the size of a wide area. Uncertainty intervenes in these data [55]. In this study, the probabilistic analysis technique was applied to quantitatively consider uncertainty and reflect it in the analysis. Such methods can be used to estimate the probability that the value of a state function satisfies a threshold by assuming that the input variable is randomly selected from a specific distribution. The probabilistic analysis method replaces the safety factor to evaluate the risk of slope using the probability of failure and is evaluated as the most effective method to quantify uncertainty among the various techniques proposed so far.

Probabilistic methods include MC simulation, first-order second moment (FOSM), the point estimate method (PEM), etc. FOSM and PEM are approximate methods, so their accuracy is relatively low compared to MC simulation [31,56]. MC simulation can represent the variability of input variables by randomly generating input variables, and is suitable for analyzing one or more random variables [57]. In this study, analysis was performed using Monte Carlo simulation among the probabilistic analysis techniques.

The MC simulation method first determines the model of the state function. The state function used in landslide susceptibility analysis depends on FS; and we calculate the probability of failure, which is the probability that FS is less than 1. Second, the probability distribution of the input variables is calculated. Because information about the probability distribution of input variables is not available, this is considered to be a normal distribution with an appropriate average and standard deviation [28,32,33,58]. Third, N random values were extracted from the distribution of input variables, and N values of FS were calculated by substituting the values of the randomly generated input variables. Finally, the probability of failure, i.e., the proportion of the N FS values that are less than 1, was calculated as follows:

$$P_f = p(FS < 1) \tag{4}$$

4.3. Applications and Validation

In our MC simulation, the cohesion and frictional angle were considered as random variables. According to previous studies, the probability distributions of geotechnical characteristics follow a normal or lognormal distribution, so the probability distribution of the input variables was assumed to be normal in this study [59–61]. Since the mean and standard deviation are required to define the distribution of input variables, the average was calculated based on the constructed data, and the standard deviation was calculated from the coefficient of variation. In previous studies, the ranges of the coefficient of variation of the internal friction angle and cohesion were 10–20% and 25–30%, respectively [56,62,63]. Because the entire study area was composed of gneiss and thus the geological characteristics were relatively similar, the minimum value of each coefficient of variation was assumed to be 10% for the internal friction angle and 25% for the cohesive coefficient, and the number of repetitions was set to 100,000.

The probability of failure is established by probabilistic methods, in contrast to the deterministic analysis in which we interpret an area to be unstable when FS is less than 1 [64]. Therefore, based on the results of previous studies, landslide-susceptible areas were classified based on failure probabilities of 1%, 5%, and 10% [64]. We then calculated the landslide prediction rate, which is the ratio of the number of landslides in a landslide-susceptible area. Furthermore, we carried out a deterministic analysis of the same dataset and calculated the FS by taking a simple average of the data. Finally, we compared the results from the MC simulation to those obtained using deterministic techniques.

Finally, the area under the curve (AUC) of the receiver operating characteristic (ROC) was calculated for validation purposes. The *x*-axis of the ROC graph is the ratio of the expected landslide area where it shows high susceptibility of landslides, and the *y*-axis is the probability of landslides. The AUC is calculated from the area under the ROC graph and has values between 0 and 1. The closer it is to 1 (100%), the higher the accuracy [45].

5. Result

The results of the MC simulation are summarized in Table 4 and shown in Figure 9. When a 1% probability of failure was considered to indicate a susceptible area, 57.75% and 42.25% of the study area was found to be unstable and stable, respectively, and the landslide prediction rate was 95.15%. When the susceptible area was set based on a probability of failure of 5%, the unstable area was 54.23%, the stable area was 45.77%, and the landslide prediction rate was 91.26%. When we defined the susceptible areas based on a probability of failure of 10%, the unstable area was 52.02%, the stable area was 47.98%, and the landslide prediction rate was 90.29%. By contrast, according to the deterministic analysis method, the unstable area was 42.72%, the stable area was 57.28%, and the prediction rate, which in this case indicated the proportion of the predicted landslides that occurred, was 81.55%. Additionally, the AUC calculation of the ROC graph of the MC simulation was 0.7332 (73.32%) (Figure 10).

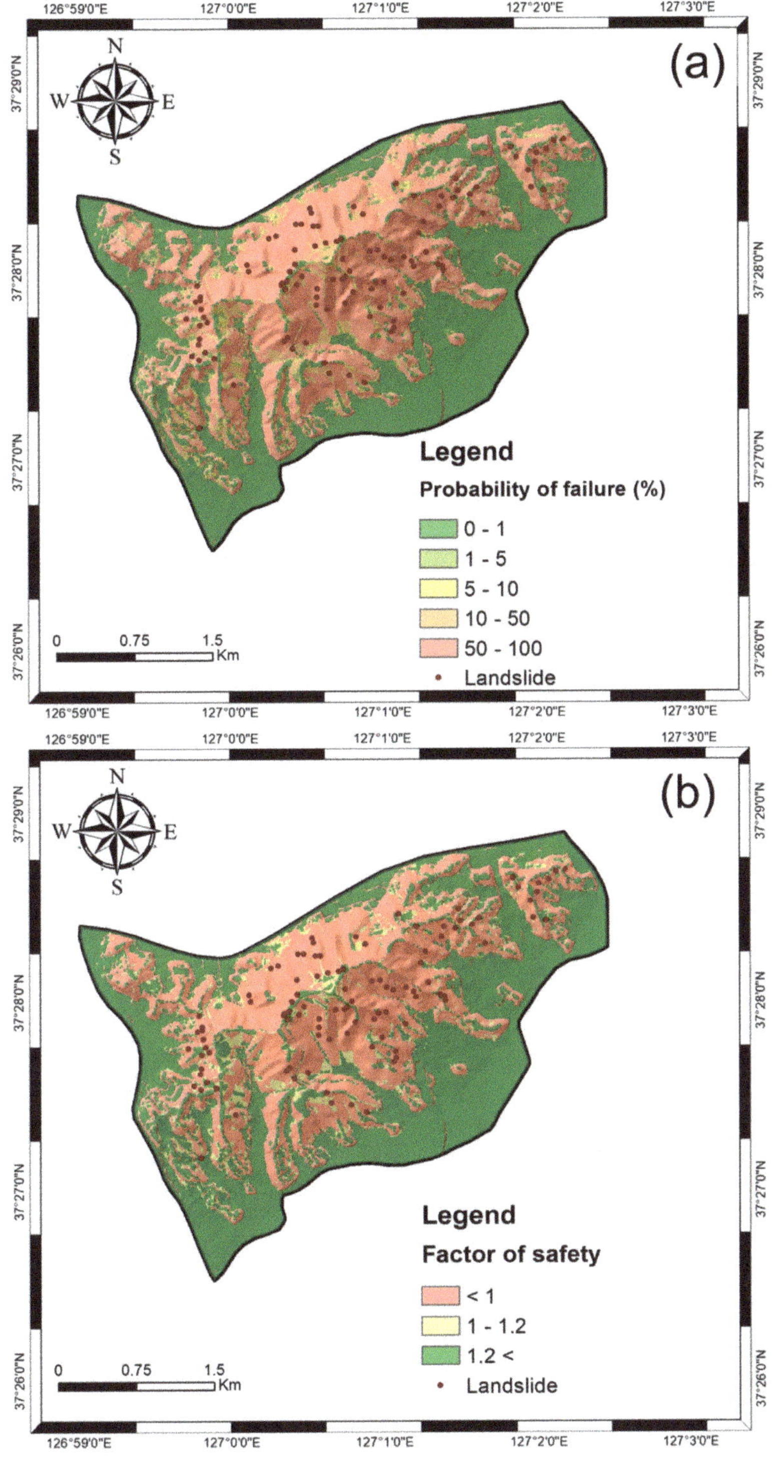

Figure 9. Map showing (**a**) the probability of failure evaluated using Monte Carlo simulation and (**b**) the factor of safety evaluated using the deterministic analysis.

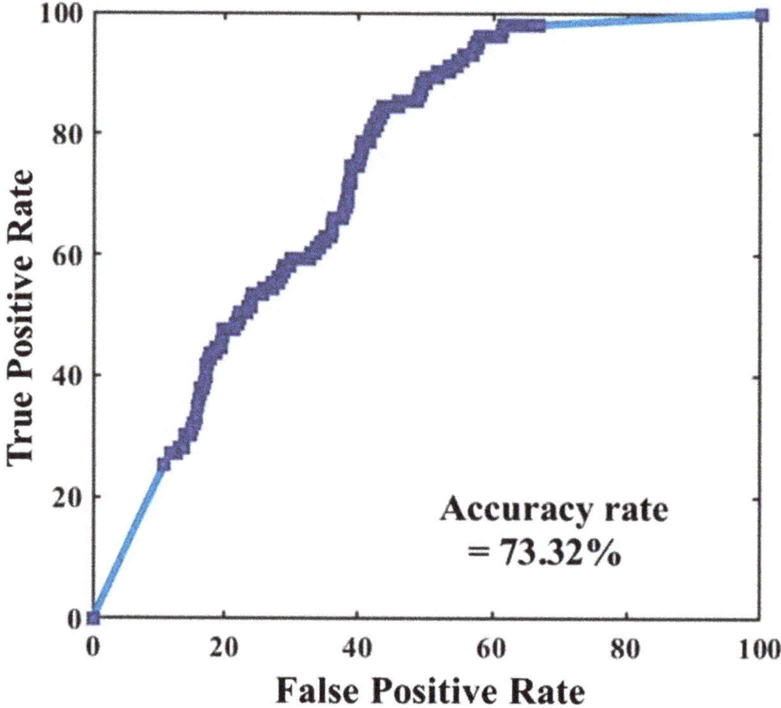

Figure 10. Receiver operating curve of the results from the Monte Carlo simulation.

Table 4. Monte Carlo simulation results.

Method	Criteria for Establishing Unstable Area	Stable Area (%)	Unstable Area (%)	Landslide Prediction Rate (%)
Monte Carlo simulation	more than 1%	42.25	57.75	95.15
Monte Carlo simulation	more than 5%	45.77	54.23	91.26
Monte Carlo simulation	more than 10%	47.98	52.02	90.29
Deterministic analysis	less than 1	57.28	42.72	81.55

Based on the results of both analytical methods, we conclude that the stable area occupies more than 40% of the study area. According to the results of the MC simulation, as the reference probability of failure decreases, the proportion of unstable regions increases with increasing landslide prediction rate. Our MC simulation predicted relatively high proportions of unstable areas and landslide prediction rates compared to the deterministic analysis. An AUC value of 0.7 or more can be interpreted as indicating good predictive performance [65,66]. The AUC obtained from the MC simulation (0.7332) was greater than 0.7, which could explain the high landslide prediction rate. The good predictive performance of the MC simulation is attributed to the variability of the input variables.

6. Discussion

In this study, we conducted a landslide susceptibility analysis of the Mt. Umyeon area, a representative example of urban landslide damage. An interpolation method was used to construct data for the entire study area because we could only obtain subsurface data and geotechnical characteristics for a few points. In addition, it was infeasible to obtain sufficient data to infer the probability distribution due to the cost and constraints of acquiring the actual property values, so we assumed the probability distribution to be normal. Since it was difficult to obtain a sufficient amount of field survey data compared to the scope of the study area, but it was difficult to apply the unsaturated soil theory, which requires many experimental values and well reflects the behavior of the groundwater level of the actual ground; saturated soil theory with relatively simple interpretation was applied in this study. In addition, several hydraulic parameters were assumed based on existing studies. Therefore, a more accurate groundwater model can be constructed with sufficient data on hydraulic parameters obtained through experiments. Future studies should increase the number of data acquisition points and experiments at each point to compensate for these limitations.

The results of the landslide susceptibility analysis of the Mt. Umyeon area can be summarized as follows. The MC simulation results showed that the landslide prediction rate (90.25, 91.26, and 95.15%) was significantly higher than the landslide prediction rate (81.55%) of the deterministic method. Comparing the results of the deterministic analysis method and the MC simulation, which is a probabilistic analysis method, by using the coincidence ratio with the location coordinate of the landslide, the landslide prediction rate of 8.7% to 13.6% is higher than the deterministic analysis method. Through these results, the probability analysis method can be applied considering the variability of the rainfall intensity, so accuracy can be improved by considering the uncertainty inherent in the rainfall intensity. MC simulation result also yielded a high AUC value (0.7332). We attributed this improvement in results to the fact that MC simulation can include the variability and uncertainty of the input variables. Moreover, the proportion of unstable areas around Mt. Umyeon exceeded 40%. This is thought to be due to the weathered gneiss distributed throughout the area and the high rainfall intensity.

7. Conclusions

Following the Mt. Umyeon landslide, a representative landslide case in the Seoul metropolitan city, South Korea, it has become necessary to analyze landslide susceptibility. The purpose of this study is to analyze the susceptibility of landslide disasters considering uncertainty before the occurrence of widespread landslides. Analysis based on physical slope models can be used to evaluate landslide susceptibility in areas without prior landslide occurrence. A spatial database of Mt. Umyeon, the study area, was constructed and analyzed by applying it to the infinite slope model that is similar to the characteristics of landslide occurrence in the study area. In addition, landslide susceptibility was applied and analyzed based on a physically based model and MC simulation, a probabilistic analysis technique considering uncertainty. An infinite slope model was used as the physically based model. In the GIS environment, it was possible to analyze landslide susceptibility in the study area, and by using an infinite slope model and using probabilistic techniques, it was possible to evaluate the landslide susceptibility as a quantitative indicator of probability of failure. Furthermore, to evaluate the accuracy of our landslide predictions, we applied a deterministic method and compared the results. Finally, the accuracy of the landslide prediction was calculated based on the AUC.

This study confirms that it is possible to evaluate high-accuracy landslide susceptibility without prior information of landslide occurrences by combining a physical slope model with probabilistic method. By varying the reference probability of failure from 1% to 10% in the MC simulation, it was possible to adjust the safety level as needed. This means that the reference failure probability can be varied according to the purpose of analysis. For example, a susceptibility map with a high standard probability of failure should be used to determine the locations of disaster prevention structures to minimize costs. Conversely, if the danger zone is temporarily set to minimize human damage, the susceptibility map should be based on the minimum probability of failure. In this way, the same dataset and probabilistic technique can be used for different purposes.

To prevent a repeat of the damage incurred by the Mt. Umyeon landslide, it is necessary to carry out landslide susceptibility studies of areas where landslides have not occurred. In particular, prior landslide susceptibility analysis should be carried out in areas with high population densities to minimize large-scale damage. The methodology presented herein can be used to prepare measures to reduce the damage caused by landslides by analyzing landslide susceptibilities in areas without landslide occurrence data, where landslides have not occurred previously. Furthermore, this methodology can be applied to various regions by extracting input factors by setting an infinite slope model that reflects regional characteristics in consideration of landslide characteristics.

Author Contributions: Conceptualization, S.L. and M.-J.L.; data curation, S.L. and J.J.; formal analysis, Y.K. and N.C.; funding acquisition, M.-J.L.; investigation, J.J. and S.L.; methodology, J.J., Y.K. and N.C.; resources, S.L. and M.-J.L.; software, J.J.; validation, J.J., Y.K. and N.C.; supervision, M.-J.L.; visualization, S.L. and J.J.; writing—original draft, J.J.; writing—review and editing, S.L. and M.-J.L. All authors have read and agreed to the published version of the manuscript.

Funding: This research was conducted at Korea Environment Institute (KEI) with support from Basic Science Research Program through the National Research Foundation of Korea (NRF) funded by the Ministry of Education (NRF-2018R1D1A1B07041203). This research was also supported by a grant (20010017) of R&D Program of Responding Technology to Climate Disaster such as Heat wave, funded by Ministry of Interior and Safety (MOIS, Korea).

Conflicts of Interest: The authors declare no conflict of interest.

References

1. Hong, H.; Shahabi, H.; Shirzadi, A.; Chen, W.; Chapi, K.; Ahmad, B.B.; Roodposhti, M.S.; Hesar, A.Y.; Tian, Y.; Bui, D.T. Landslide susceptibility assessment at the Wuning area, China: A comparison between multi-criteria decision making, bivariate statistical and machine learning methods. *Nat. Hazards* **2019**, *96*, 173–212. [CrossRef]
2. Zhao, X.; Chen, W. Optimization of Computational Intelligence Models for Landslide Susceptibility Evaluation. *Remote Sens.* **2020**, *12*, 2180. [CrossRef]
3. Lee, S.; Lee, M.-J.; Jung, H.-S.; Lee, S. Landslide susceptibility mapping using naïve bayes and bayesian network models in Umyeonsan, Korea. *Geocarto Int.* **2019**, 1–15. [CrossRef]
4. Pham, B.T.; Prakash, I.; Singh, S.K.; Shirzadi, A.; Shahabi, H.; Bui, D.T. Landslide susceptibility modeling using Reduced Error Pruning Trees and different ensemble techniques: Hybrid machine learning approaches. *Catena* **2019**, *175*, 203–218. [CrossRef]
5. Son, J.; Suh, J.; Park, H.-D. GIS-based landslide susceptibility assessment in Seoul, South Korea, applying the radius of influence to frequency ratio analysis. *Environ. Earth Sci.* **2016**, *75*, 310. [CrossRef]
6. Akgun, A.; Türk, N. Landslide susceptibility mapping for Ayvalik (Western Turkey) and its vicinity by multicriteria decision analysis. *Environ. Earth Sci.* **2010**, *61*, 595–611. [CrossRef]
7. Lee, S.; Lee, M.-J.; Jung, H.-S. Data mining approaches for landslide susceptibility mapping in Umyeonsan, Seoul, South Korea. *Appl. Sci.* **2017**, *7*, 683. [CrossRef]
8. Park, S.; Choi, C.; Kim, B.; Kim, J. Landslide susceptibility mapping using frequency ratio, analytic hierarchy process, logistic regression, and artificial neural network methods at the Inje area, Korea. *Environ. Earth Sci.* **2013**, *68*, 1443–1464. [CrossRef]
9. Pradhan, A.M.S.; Kim, Y.-T. Spatial data analysis and application of evidential belief functions to shallow landslide susceptibility mapping at Mt. Umyeon, Seoul, Korea. *Bull. Eng. Geol. Environ.* **2017**, *76*, 1263–1279. [CrossRef]
10. Bui, D.T.; Tsangaratos, P.; Nguyen, V.-T.; Van Liem, N.; Trinh, P.T. Comparing the prediction performance of a Deep Learning Neural Network model with conventional machine learning models in landslide susceptibility assessment. *Catena* **2020**, *188*, 104426. [CrossRef]
11. Kavzoglu, T.; Colkesen, I.; Sahin, E.K. Machine Learning Techniques in Landslide Susceptibility Mapping: A Survey and a Case Study. In *Landslides: Theory, Practice and Modelling*; Springer: Berlin/Heidelberg, Germany, 2019; pp. 283–301.
12. Dou, J.; Yunus, A.P.; Bui, D.T.; Merghadi, A.; Sahana, M.; Zhu, Z.; Chen, C.-W.; Khosravi, K.; Yang, Y.; Pham, B.T. Assessment of advanced random forest and decision tree algorithms for modeling rainfall-induced landslide susceptibility in the Izu-Oshima Volcanic Island, Japan. *Sci. Total Environ.* **2019**, *662*, 332–346. [CrossRef]
13. Park, S.; Kim, J. Landslide susceptibility mapping based on random forest and boosted regression tree models, and a comparison of their performance. *Appl. Sci.* **2019**, *9*, 942. [CrossRef]

14. Sevgen, E.; Kocaman, S.; Nefeslioglu, H.A.; Gokceoglu, C. A novel performance assessment approach using photogrammetric techniques for landslide susceptibility mapping with logistic regression, ANN and random forest. *Sensors* **2019**, *19*, 3940. [CrossRef]

15. Abedini, M.; Ghasemian, B.; Shirzadi, A.; Bui, D.T. A comparative study of support vector machine and logistic model tree classifiers for shallow landslide susceptibility modeling. *Environ. Earth Sci.* **2019**, *78*, 560. [CrossRef]

16. Shahri, A.A.; Spross, J.; Johansson, F.; Larsson, S. Landslide susceptibility hazard map in southwest Sweden using artificial neural network. *Catena* **2019**, *183*, 104225. [CrossRef]

17. Tian, Y.; Xu, C.; Hong, H.; Zhou, Q.; Wang, D. Mapping earthquake-triggered landslide susceptibility by use of artificial neural network (ANN) models: An example of the 2013 Minxian (China) Mw 5.9 event. *Geomat. Nat. Hazards Risk* **2019**, *10*, 1–25. [CrossRef]

18. Korea Forest Service. *Detailed Strategy for Primary Policy*; Korea Forest Service: Seoul, Korea, 2013.

19. Burton, A.; Bathurst, J. Physically based modelling of shallow landslide sediment yield at a catchment scale. *Environ. Geol.* **1998**, *35*, 89–99. [CrossRef]

20. Goetz, J.N.; Guthrie, R.H.; Brenning, A. Integrating physical and empirical landslide susceptibility models using generalized additive models. *Geomorphology* **2011**, *129*, 376–386. [CrossRef]

21. Abbott, M.B.; Bathurst, J.C.; Cunge, J.A.; O'Connell, P.E.; Rasmussen, J. An introduction to the European Hydrological System—Systeme Hydrologique Europeen, "SHE", 1: History and philosophy of a physically-based, distributed modelling system. *J. Hydrol.* **1986**, *87*, 45–59. [CrossRef]

22. Flentje, P.N.; Miner, A.; Whitt, G.; Fell, R. Guidelines for landslide susceptibility, hazard and risk zoning for land use planning. *Eng. Geol.* **2007**, *102*, 3–4.

23. Ho, J.-Y.; Lee, K.T.; Chang, T.-C.; Wang, Z.-Y.; Liao, Y.-H. Influences of spatial distribution of soil thickness on shallow landslide prediction. *Eng. Geol.* **2012**, *124*, 38–46. [CrossRef]

24. Santoso, A.M.; Phoon, K.-K.; Quek, S.-T. Effects of soil spatial variability on rainfall-induced landslides. *Comput. Struct.* **2011**, *89*, 893–900. [CrossRef]

25. Einstein, H.; Baecher, G. Probabilistic and Statistical Methods in Engineering Geology I. Problem Statement and Introduction to Solution. In *Ingenieurgeologie und Geomechanik als Grundlagen des Felsbaues/Engineering Geology and Geomechanics as Fundamentals of Rock Engineering*; Springer: Berlin/Heidelberg, Germany, 1982; pp. 47–61.

26. El-Ramly, H.; Morgenstern, N.; Cruden, D. Probabilistic slope stability analysis for practice. *Can. Geotech. J.* **2002**, *39*, 665–683. [CrossRef]

27. Hanss, M.; Turrin, S. A fuzzy-based approach to comprehensive modeling and analysis of systems with epistemic uncertainties. *Struct. Saf.* **2010**, *32*, 433–441. [CrossRef]

28. Mostyn, G.; Li, K. Probabilistic Slope Analysis-State-of-Play. In Proceedings of the Conference on Probabilistic Methods in Geotechnical Engineering, Canberra, Australia, 10–12 February 1993; pp. 89–109.

29. Mostyn, G.R.; Small, J. Methods of stability analysis. In *Soil Slope Instability and Stabilization*, 1st ed.; Walker, B.F., Fell, R., Eds.; Balkema: Rotterdam, The Netherlands, 1987; pp. 71–120.

30. Nilsen, B. New trends in rock slope stability analyses. *Bull. Eng. Geol. Environ.* **2000**, *58*, 173–178. [CrossRef]

31. Park, H.; West, T. Development of a probabilistic approach for rock wedge failure. *Eng. Geol.* **2001**, *59*, 233–251. [CrossRef]

32. Park, H.-J.; West, T.R.; Woo, I. Probabilistic analysis of rock slope stability and random properties of discontinuity parameters, Interstate Highway 40, Western North Carolina, USA. *Eng. Geol.* **2005**, *79*, 230–250. [CrossRef]

33. Pathak, S.; Nilsen, B. Probabilistic rock slope stability analysis for Himalayan conditions. *Bull. Eng. Geol. Environ.* **2004**, *63*, 25–32. [CrossRef]

34. Lee, J.H.; Park, H.J. Assessment of landslide susceptibility using a coupled infinite slope model and hydrologic model in Jinbu area, Gangwon-do. *Econ. Environ. Geol.* **2012**, *45*, 697–707. [CrossRef]

35. Beven, K.J.; Kirkby, M.J. A physically based, variable contributing area model of basin hydrology/Un modèle à base physique de zone d'appel variable de l'hydrologie du bassin versant. *Hydrol. Sci. J.* **1979**, *24*, 43–69. [CrossRef]

36. Montgomery, D.R.; Dietrich, W.E. A physically based model for the topographic control on shallow landsliding. *Water Resour. Res.* **1994**, *30*, 1153–1171. [CrossRef]

37. Korean Society of Civil Engineers. *Complementary Studies on the Cause of the Umyeonsan(Mt.)*; Korean Society of Civil Engineers: Seoul, Korea, 2012.

38. Korean Geotechnical Society. *Final Report on the Cause of Landslides in Umyeonsan(Mt.) Area and the Establishment of Restoration Measures*; Korean Geotechnical Society: Seoul, Korea, 2012.

39. All in Korea. What Are the Causes of Umyeonsan Landslide? Available online: http://www.allinkorea.net/22313 (accessed on 1 August 2020).

40. MBC News. Causes of Umyeonsan Landslide, Other than '120 Years of Heavy Rain'. Availabe online: http://d.kbs.co.kr/news/view.do?ncd=2825925 (accessed on 1 August 2020).

41. Jang, J.Y.; Park, H.J. Physically Based Landslide Susceptibility Analysis Using a Fuzzy Monte Carlo Simulation in Sangju Area, Gyeongsangbuk-Do. *Econ. Environ. Geol.* **2017**, *50*, 239–250.

42. Jeong, S.; Kim, Y.; Lee, J.K.; Kim, J. The 27 July 2011 debris flows at Umyeonsan, Seoul, Korea. *Landslides* **2015**, *12*, 799–813. [CrossRef]

43. Kim, S.; Paik, J.; Kim, K.S. Run-out modeling of debris flows in Mt. Umyeon using FLO-2D. *J. Korean Soc. Civ. Eng.* **2013**, *33*, 965–974. [CrossRef]

44. Lee, G.-H.; Oh, S.-R.; Lee, D.-U.; Jung, K.-S. Analysis on Mt. Umyeon Landslide Using Infinite Slope Stability Model. In Proceedings of the Korea Water Resources Association Conference, Jeongseon, Gangwon, Korea, 17–18 May 2012; pp. 737–741.

45. Lee, S.; Lee, M.-J. Susceptibility mapping of Umyeonsan using logistic regression (LR) model and post-validation through field investigation. *Korean J. Remote Sens.* **2017**, *33*, 1047–1060.

46. Pradhan, A.M.S.; Kang, H.-S.; Lee, J.-S.; Kim, Y.-T. An ensemble landslide hazard model incorporating rainfall threshold for Mt. Umyeon, South Korea. *Bull. Eng. Geol. Environ.* **2019**, *78*, 131–146. [CrossRef]

47. Kakaomap. Kakaomap. Available online: https://map.kakao.com/ (accessed on 1 July 2020).

48. Saulnier, G.-M.; Beven, K.; Obled, C. Including spatially variable effective soil depths in TOPMODEL. *J. Hydrol.* **1997**, *202*, 158–172. [CrossRef]

49. Crosta, G.B.; Frattini, P. Rainfall-induced landslides and debris flows. *Hydrol. Process. Int. J.* **2008**, *22*, 473–477. [CrossRef]

50. Murphy, W.; Vita-Finzi, C. Landslides and seismicity-An application of remote sensing. In Proceedings of the Thematic Conference on Geologic Remote Sensing, Denver, CO, USA, 29 April–2 May 1991; pp. 771–784.

51. Ward, T.J.; Li, R.-M.; Simons, D.B. Mapping landslide hazards in forest watersheds. *J. Geotech. Geoenviron. Eng.* **1982**, *108*, 319–324.

52. Van Westen, C.; Terlien, M. An approach towards deterministic landslide hazard analysis in GIS. A case study from Manizales (Colombia). *Earth Surf. Process. Landf.* **1996**, *21*, 853–868. [CrossRef]

53. Coduto, D.P. *Geotechnical Engineering: Principles and Practices*; Prentice Hall: Upper Saddle River, NJ, USA, 1999.

54. O'loughlin, E. Prediction of surface saturation zones in natural catchments by topographic analysis. *Water Resour. Res.* **1986**, *22*, 794–804. [CrossRef]

55. You, K.; Park, Y.; Lee, J.S.J.T.; Technology, U.S. Risk analysis for determination of a tunnel support pattern. *Int. J. Adv. Struct. Eng.* **2005**, *20*, 479–486. [CrossRef]

56. Harr, M.E. *Reliability-Based Design in Civil Engineering*; Dover Publications: Mineola, NY, USA, 1984.

57. Greco, V.R. Efficient Monte Carlo technique for locating critical slip surface. *J. Geotech. Eng.* **1996**, *122*, 517–525. [CrossRef]

58. Park, H.-J. The Evaluation of Failure Probability for Rock Slope Based on Fuzzy Set Theory and Monte Carlo Simulation. *J. Korean Geotech. Soc.* **2007**, *23*, 109–117.

59. Liu, C.-N.; Wu, C.-C. Mapping susceptibility of rainfall-triggered shallow landslides using a probabilistic approach. *Environ. Geol.* **2008**, *55*, 907–915. [CrossRef]

60. Lumb, P. The variability of natural soils. *Can. Geotech. J.* **1966**, *3*, 74–97. [CrossRef]

61. Oka, Y.; Wu, T.H. System reliability of slope stability. *J. Geotech. Eng.* **1990**, *116*, 1185–1189. [CrossRef]

62. Luo, Z.; Atamturktur, S.; Juang, C.H.; Huang, H.; Lin, P.-S. Probability of serviceability failure in a braced excavation in a spatially random field: Fuzzy finite element approach. *Comput. Geotech.* **2011**, *38*, 1031–1040. [CrossRef]

63. Phoon, K.-K.; Retief, J.V. *Reliability of Geotechnical Structures in ISO2394*; CRC Press: Boca Raton, FL, USA, 2016.

64. Chowdhury, R.; Flentje, P.; Bhattacharya, G. *Geotechnical Slope Analysis*; CRC Press: Boca Raton, FL, USA, 2009.

65. Fawcett, T. An introduction to ROC analysis. *Pattern Recognit. Lett.* **2006**, *27*, 861–874. [CrossRef]
66. Lee, S.; Hyun, Y.; Lee, S.; Lee, M.-J. Groundwater Potential Mapping Using Remote Sensing and GIS-Based Machine Learning Techniques. *Remote Sens.* **2020**, *12*, 1200. [CrossRef]

Letter

Intelligent WSN System for Water Quality Analysis Using Machine Learning Algorithms: A Case Study (Tahuando River from Ecuador)

Paul D. Rosero-Montalvo [1,2,*], Vivian F. López-Batista [1], Jaime A. Riascos [3,4] and Diego H. Peluffo-Ordóñez [4,5,6]

1. Department of Computer Science and Automatics Salamanca, Universidad de Salamanca, 37008 Salamanca, Spain; vivian@usal.es
2. Department of Applied Sciences, Universidad Técnica del Norte, 100150 Ibarra, Ecuador
3. Department of Engineering, Universidad Mariana, 520001 Pasto, Colombia; jandresr@umariana.edu.co
4. Department of Engineering, Corporación Universitaria Autónoma de Nariño, 520002 Pasto, Colombia; dpeluffo@yachaytech.edu.ec
5. School of Mathematical and Computational Sciences, Universidad Yachay Tech, 100650 Urcuquí, Ecuador
6. SDAS Researh Group, 100150 Ibarra, Ecuador
* Correspondence: pdrosero@utn.edu.ec

Received: 24 April 2020; Accepted: 12 June 2020; Published: 20 June 2020

Abstract: This work presents a wireless sensor network (WSN) system able to determine the water quality of rivers. Particularly, we consider the Tahuando River from Ibarra, Ecuador, as a case study. The main goal of this research is to determine the river's status throughout its route, by generating data reports into an interactive user interface. To this end, we use an array of sensors collecting several measures such as: turbidity, temperature, water quality, pH, and temperature. Subsequently, from the information collected on an Internet-of-Things (IoT) server, we develop a data analysis scheme with both data representation and supervised classification. As an important result, our system outputs a map that shows the contamination levels of the river at different regions. Furthermore, in terms of data analysis performance, the proposed system reduces the data matrix by 97% from its original size, while it reaches a classification performance over 90%. Furthermore, as an additional remarkable result, we here introduce the so-called quantitative metric of balance (QMB), which measures the balance or ratio between performance and power consumption.

Keywords: prototype selection; river pollution; supervised classification; WSN

---

1. Introduction

Rivers are natural watercourses that commonly come from both precipitation (surface runoff), and snowpacks (e.g., water stored in glaciers). Regularly, they flow towards lakes, sea, oceans, or another river. Urban rivers are responsible for providing water resources to crops and human beings as well as navigation purposes. Certainly, this natural resource may not be everlasting. As a matter of fact, there is currently a great deficit of water reserves due to deforestation, inappropriate and excessive use of fertilizers and pesticides, causing environmental issues [1,2]. Likewise, the urbanization and industries have had collateral adverse impact directly on the water quality of river ecosystems worldwide [3]. Besides, the population growth produces enormous wastewater that enters into the rivers without any environmental control. United Nations (UN) settled that 90% of such waste is not correctly treated, and 70% of the industries discharge contaminant content without any adequate standards or rigorous inspections [4,5]. Water pollution contains high levels of biochemical oxygen demand (BOD), nitrogen, and phosphorus. So it is necessary to develop systems that support the detection and measuring of the contamination levels in rivers to maintain an optimal ecological balance, limiting environmental damage and preventing diseases spread [6]. Consequently, city governments have stated environmental policies intended to create urban regeneration initiatives around the care of their rivers [7]. In this connection, Ecuador, as our case study, has no any short or long-term plan to improve either the urban or rural river conditions [8].

Traditionally, water quality monitoring uses collected samples for laboratory testing, enabling then a wide range of analyses. Notwithstanding, it results impractical to manually measure water pollution at different points along the river. Moreover, this sort of tests may take a few days, and probably not reaching as a good precision as that of in-situ sampling [9]. Nowadays, the use of sensors for monitoring environmental conditions has received significant attention due to the real-time data collection, flexibility, and portability [4,10]. Following from this, the creation of a wireless sensor network (WSN) that combines several sensors with a data processing system and wireless communication can allow for an adequate measure of the water quality, where each sensor becomes a node that shares information among them as well as to a central server [11,12]. Thus, these data are greatly useful for further robust analyzes of water pollution in rivers. However, the large amount of data demands the implementation of machine learning algorithms to create systems that automatically can detect high levels of water pollution and make proper decisions. For that purpose, historical data (training data) become valuable to turn WSN nodes into intelligent systems [13,14].

Consequently, this work presents a novel system composed of three WSN nodes for monitoring in real-time the water pollution present in the Tahuando River (located in Ibarra, Ecuador) using machine learning algorithms. To do so, we establish different measurement points wherein each WSN node acquires the river's conditions data to be later processed internally by the system. In this sense, we consider water-quality variables, namely pH, turbidity, temperature and dissolved solids. Additionally, we carry out a sensor integration and calibration stage for eliminating reading errors. Finally, we sent these data to a cloud server, using a mobile network, where we visualize the node's information with its proper geo-location. As relevant results, a reduction of the required training set of 97% is accomplished by using is the condensed nearest neighbor (CNN) method as a prototype selection approach, as well as the classification stage—with k-NN—reaches 90.6% of performance. Then, our work is an exploratory study on different methods for both prototype selection and data classification applied to water treatment. Therefore, we have no gold standard result or benchmark method. Instead, an exhaustive comparison of representative methods is presented.

The fact that the data analysis process is implemented directly into the WSN represents a novelty itself for the development of both intelligent embedded systems, and data analysis platforms under low-computational resources. The rationale of creating an intelligent system including in-situ data analysis tasls (e.g., data classification) lies in the fact that an embedded systems can perform automatic decision-making processes with no requiring an external server. As well, it enables the possibility that even non-expert operators can readily interact with the system. In addition, it represents a solution to one of the main open issues of WSNs design, namely: information redundancy, which constraints the battery life-time, and often requires the incorporation of an external server for decision-making procedures. Additionally, to display a report of the current river's status, we implement an interactive user interface.

The rest of the manuscript is organized as follows: Section 2 gathers some remarkable related works. Section 3 describes both the system design and the data analysis proposed for implementing the machine learning algorithms. Section 4 presents the tests and results. Finally, Section 5 gathers the concluding remarks.

2. Related Works

Some works [5,6,9,15] have extensively worked on the estimation of water pollution, presenting different solutions for determining pollution state and its levels along several rivers located in China. Other works [16,17] analyze river status using satellite photographs. Meanwhile, in [4,10] WSN are instead preferred for data acquisition.

The work presented in [18] develops a WSN to determine the water quality level for human consumption through GPRS-generated data analysis, which is carried out on an external-to-WSN server holding a communication module. Similarly, another work [19] uses a high-performance external server. Specifically, it presents a system able to measure the quality of the water stored in tanks or reservoirs. In this connection, other works have proposed alternatives to improve the data processing aimed at reaching an admissible performance while involving a lower computational burden. An approach to do so is by minimizing the communication load, as done in [20] wherein an additional data compression stage is incorporated—particularly, the principal component analysis (PCA) algorithm is used. By compressing (or reducing the dimensionality of) data, the sending-packets process through WSN is enhanced in terms of performance and processing time. Similarly, the work presented in [21] performs a data analysis including temperature, pH, electrical conductivity (EC) and dissolved oxygen (DO) sensors, whose data are processed on a server and its result is sent back to the proposed WSN for decision-making. Another approach, which is becoming a new embedded systems paradigm is the design of intelligent systems performing an in-situ data analysis. For instance, in [22] the redundancy is minimized following a data fusion criterion to better manage the WSN computational resources, and bring an adequate energy consumption. Under this new paradigm where data analysis is carried out into the same system handling the data acquisition, the design of a system related to water quality monitoring results not only novel but proper. Indeed, on doing so, there would be enabled an affordable, large-coverage and easy-to-use WSN system, which along with right sensors will help environmental or health-related agencies or bodies to effectively make decisions regarding the quality of natural water from a specific source. Following from this, the work [22] involves stages for data acquisition, processing, and visualization.

Nonetheless, no one of these solutions presents an in-situ data analysis. From the reviewed literature, only [23] presents an analysis of rivers in Ecuador. All of the aforementioned works presented appealing solutions to determine the water conditions of different rivers. However, in spite of all these efforts, there are still many open issues, such as: real-time data analysis, sensor calibration, and sending information to storage servers located far away from the acquisition point, among others.

3. Materials and Methods

Broadly, the proposed system consists of the following stages: initial conditions of the study region (Section 3.1), WSN design for accurate data acquisition (Section 3.2), and the data analysis with both the criteria for prototype selection, and supervised classification (Section 3.3).

3.1. Initial Conditions of the Study Region

The city of Ibarra (Ecuador) is the capital of the province of Imbabura with a dry-temperate climate of 18 °C on average. The urban population is 109 thousand and a rural population of approximately 45 thousand inhabitants. Its main commercial activity is the production of wooden articles and services to medium-scale industries. Regarding its water supply, 90% is carried out through the public distribution network, while the rest is for the use of river and vertier water [24]. Tahuando river is an important water resource in the Imbabura province, being part of the natural system of Ecuador. Due to its ability to transport and the flowing of its waters, it can withstand a large number of pollutants. However, there are several modifications at the ecological level, such as the loss of aquatic species, foul-smelling, and watercolor changes, among others. In Ecuador, only 10% of wastewater is treated. In Ibarra, around 600 liters per second of these waters are discharged into the Tahuando River, causing that no urban regeneration based on the increase in tourism can be carried out [25]. The Tahuando River is located at 0.4° latitude and 78.13° longitude. It encompasses an extension of 12 km from the community of Pesillo towards Salinas, in the Ibarra city. Figure 1 depicts the geographical location and basin.

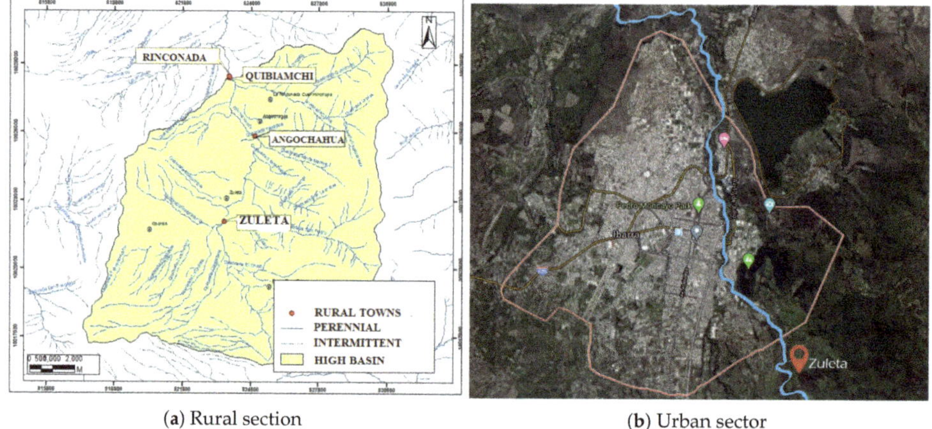

(a) Rural section   (b) Urban sector

Figure 1. Geographical description of Tahuando river. Zoomed view of the river route highlighting remarkable surrounding communities in Ibarra city's urban sector (**right**), and a widespread view regarding the rural section of Imbabura province (**left**).

3.2. Wireless-Sensor-Network Design

The design of our WSN approach is followed from the considered water-quality-related variables: pH, turbidity, temperature and dissolved solids. The considered sensor network is as follows: Firstly, we measure the turbidity and identify what kind of pollutants can be found in the river, such as: wastewater, chemicals, among others. Secondly, we use a pH sensor to determine if the water composition is acidic or basic as well as a quality sensor (total dissolved solids, TDS) to assess the level of dissolved oxygen in the water (cleanliness) [9]. Thirdly, we incorporate a temperature sensor to determine the water's changes and its relationship with the rest of the variables. To suitably develop the WSN network, we consider several operational requirements in the selection of sensors, such as reliability, precision, availability, ease-of-use, and scalability. Furthermore, in the selection of the WSN network processor system, we consider the number of pins and sensor libraries, as done in a previous work [11]. Specifically, the considered sensors are: SKU: SEN0189 (turbidity), SKU: PH-7BNC (pH), Ds18b20 (temperature), RB-Dfr-797 (TDS). As well, the Arduino Uno is selected as processing system. Additionally, we use both global position module (GPS) and mobile communications (GSM) Sim808 to send data. Finally, there is a Lipo rider battery manager for power supply with a solar charging system. Figure 2 presents the considered sensors along with the processor system (Arduino Uno).

Likewise, we calibrate each sensor as follows: sensor SKU:PH-7BNC (pH) has a linear response, so its tuning is based on measuring the voltage of several pH solutions. Particularly, we use two solutions, the first one was pH = 4.01, getting a voltage of 2.98 volts; meantime, the second one was pH = 6.86, obtaining a voltage of 2.53 v. Thus, the equation to obtain the estimated pH is:

$$\text{pH} = -5.65 * (v_1) + 21.15, \tag{1}$$

with v_1 as the voltage obtained by the sensor SKU:PH-7BNC. Likewise, the turbidity sensor SKU: SEN0189 gives a reading ranging between 2.5 to 4.3 volts with values between 3000 and 0 turbidities (NTU), respectively. According to its datasheets, we can write the following equation:

$$\text{NTU} = -1120.4 * v_2{}^2 + 5742.3 * v_2 - 4352.9, \tag{2}$$

where v_2 is the voltage registered by the sensor SKU:SENO189. On the other hand, the datasheet of the temperature sensor Ds18b20 indicates that each Celsius degree can be transformed using the equality $10\,\text{mv} = 1\,^{\circ}\text{C}$; thus, the equation is:

$$Temp = \frac{v_3 * 5}{1023 * 0.01}, \tag{3}$$

with v_3 as the voltage obtained by the sensor Ds18b20.

Finally, the TDS RB-Dfr-797 sensor provides a flexible calibration protocol, with a reset button, we can return to the initial conditions, that is, a TDS value of 23 mv. Consequently, we refresh the Arduino program and use the next equation:

$$TDS = \frac{(30 * 5 * 1000) - (75 * v_4) * 5 * (1000/1024)}{75 - 0.23}, \tag{4}$$

where v_4 is the voltage obtained by the sensor RB-Dfr-797.

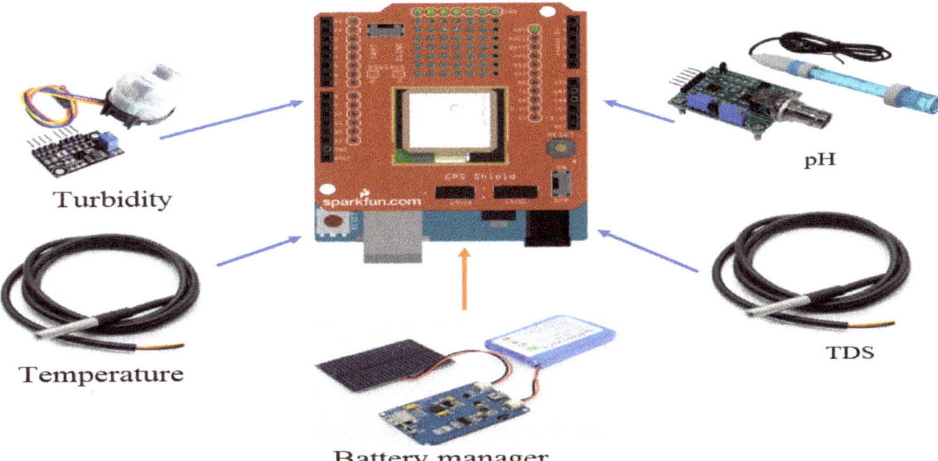

Figure 2. Demonstrative diagram of the proposed WSN system. Considered sensors (SKU: SENO189 (turbidity), SKU: PH-7BNC (pH), Ds18b20 (temperature), RB-Dfr-797 (TDS)), and the processor (Arduino Uno).

Upon sensor configuration, each v_i value will correspond to a digital-analog converter (DAC) with a resolution of 10 bits, already in the microprocessor Arduino Uno. Furthermore, we implement the moving average recursive filter to reduce the acquisition errors and smoothing the signal from each DAC. This filter takes a subset (window) of N samples, and calculate its arithmetic average to estimate a filtered sample as [26]. This filter is implemented in each DAC separately through the following equation:

$$y_n = (2n + 1)^{-1} \sum_{i=n-d}^{n+d} x_i, \tag{5}$$

where $\mathbf{x} = (x_1, \ldots, x_{L_x})$ is the input signal, $\mathbf{y} = (y_1, \ldots, y_{L_y})$ is the filtered signal, d is the window size, and L_x and L_y are respectively the input and filtered signal lengths. To accounting for a reduction of the computational resources usage, we experimentally define $d = 11$.

With the aim of verifying the data obtained by each sensor and validating the reliability thereof, samples obtained from the river are taken to the Environment Services Laboratory of the Technical

University of the North (Universidad Técnica del Norte (Universidad Técnica del Norte official web site: https://www.utn.edu.ec/web/uniportal/) from Ibarra-Ecuador, as they count on the technology and reagents to make comparison against the data obtained by the WSN. In this sense, following reliability criteria for each sensor, some recommended performance measures are considered, such as: (i) Accuracy: ability to provide the same reading by repeatedly performing the same experiment (standard deviation), (ii) Reproducibility: ability to reproduce the same results when modifying initial conditions of the experiment, and (iii) Stability: ability to produce the same output value in a long time. Overall obtained results are gathered in Table 1, which correspond to 10 tests over controlled environments to assess the data stability. As can be appreciated, the collected data from the sensors exhibit an error average of 5% in contrast to the those generated at the laboratory—such an error is acceptable enough for implementation purposes.

Table 1. Sensor performance metrics.

	Sensors			
Measure	SEN0189 (turbidity)	PH-7BNC (pH)	Ds18b20 (temperature)	RB-Dfr-797 (TDS)
Precision	7 ±	3 ±	5 ±	5 ±
Reproducibility	It is necessary to wait up 2 s for calibration to be done	Adequate	Adequate	Some reading errors
Stability	Adequate	3 ±, variable for each test	Adequate	Adequate

3.3. Data Analysis Paradigm

For a proper and wide data acquisition, we establish three node points in different locations, based on the population density of Ibarra, as follows: (i) *La Rinconada*, with low population and located at the river's beginning; (ii) *El Tejar*, with middle population rate and some wastewater discharged into the river; and (iii) *La Victoria*, with a larger population density and more discharge of pollutants from the city. Figure 3 shows the geographic locations of the nodes. Furthermore, we label each data from the nodes with a localization tag. For the data acquisition procedure, we design a collection protocol as follows: A schedule consisting in four collecting times is set, namely: in the morning, afternoon, night and early morning. Such a schedule is timed with Timer2, which is an Arduino internal clock. So, the system is timed for alerts at 08:00, 12:00, 17:00 and 00:00. On those times, the system records the sensor readings every 10 min for two hours (amounting to 6 samples per hour). Finally, these captured data are sent to the remote server through the GSM/GPS sensor. This collection protocol was performed during 3 months, generating an enough amount of information to be used in the subsequent data analysis stage.

Once the data are stored in an external server, a two-stages data analysis process is carried out: The first stage is the training set size reduction—via prototype selection—involving the least or no affectation to the intrinsic knowledge they hold. The second one is the classification task, in which the the algorithm that best fits the first stage while keeping a high accuracy is sought. Both stages are set and performed under low-computational cost criteria (given the device conditions). This process is carried out in order to be compiled within each WSN node (including both prototype selection, and classification). Then, system is able to make their own decisions based upon the reduced, stored dataset as well as the implemented classification algorithm. Therefore, on the one hand, the adaptability criterion required by an intelligent system is met, by making it able to be used anywhere on the river. On the other hand, the resulting system requires no re-run the data analysis process and thus it can be readily used by any system operator whom is not required to hold an expertise on embedded systems or data analysis, but only knowledge in water treatment itself.

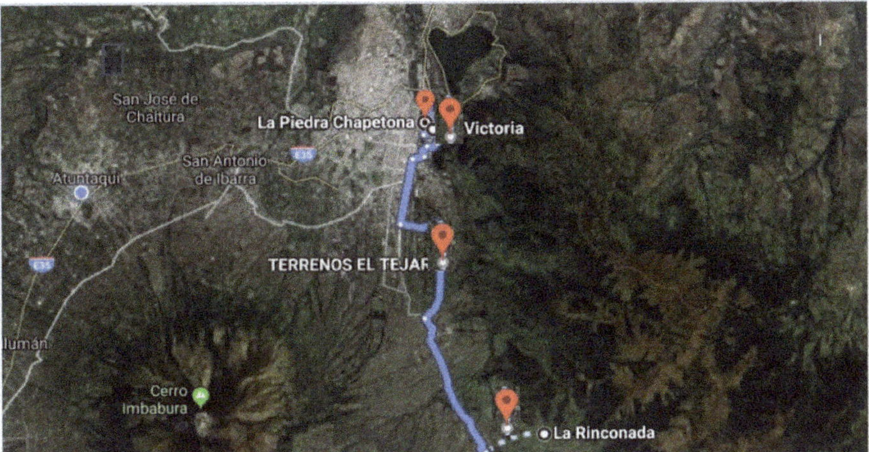

Figure 3. Geographic location of the WSN nodes. Spots strategically selected to acquire data from, in order to encompass representative zones, as well as different types and levels of river pollution.

3.3.1. Proposed Quality Measure: Quantitative Metric of Balance (QMB)

Algorithm analysis is an important part of designing thereof. Traditionally, the analysis of programming code or algorithms lies in applying theoretical and mathematical procedures. Indeed, when selecting supervised classification algorithms, efficient programs must be ensured to be created, as this translates into better power consumption and therefore battery life time usage. In this sense, the here-introduced Quantitative metric of balance (QMB) is aimed at quantifying how proper is the ratio between classifier performance and the data size reduction by the prototype selection stage. In this connection, the closer its value is to 100%, the better the ratio. As these three individual measures have an increasing nature, we multiply them to state a single value, namely, the rate of removed instances (RI) times the classification performance (CP), and divided by the response time of the classification algorithms (RT), as follows:

$$QMB = \frac{(RI * CP)}{RT} * 100\%. \tag{6}$$

Certainly, some classification criteria make use of mathematical functions or recursive functions of model adjustment that, when coded in a low-level language (assembler), generate response time delays, memory saturation and an excessive battery consumption. In this sense, the proposed QMB is aimed at penalizing the excessive computational cost in order to make it more feasible the implementation of data analysis algorithms into an embedded system. Besides, since it takes into consideration the number of removed training set instances to quantify the overall performance, this metric rewards the classification algorithm if it requires the least memory capacity when performing the decision-making procedures. When operating under real conditions, the system acquires the data from the sensors, filter the acquisition errors, make the decision through its compiled classification algorithm, and use the selection of prototypes to determine if this new reading improves the prediction ability of the system. If so, it is added into the training matrix otherwise it is only sent to the external server for visualization purposes.

Figure 4 shows the proposed data analysis scheme.

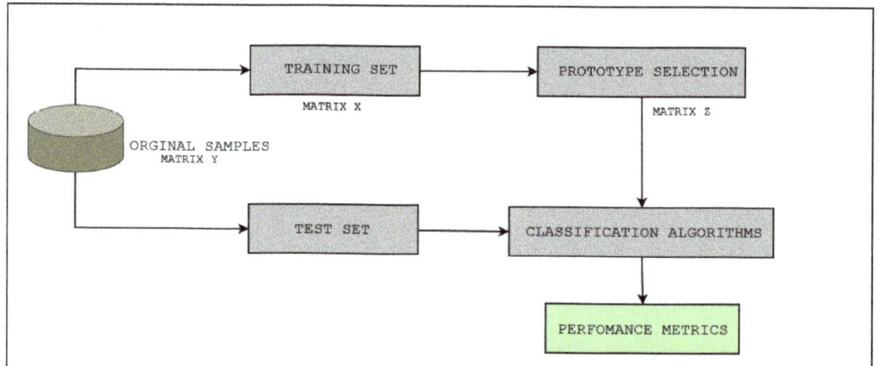

Figure 4. Data analysis scheme including prototype selection and classification stages.

3.3.2. Prototype Selection

Since WSN systems have limited computational resources, its battery consumption is directly related to the amount of data to be processed, and therefore the implementation of machine learning algorithms into thereof is limited. In this connection, the prototype selection (PS) techniques may take place by reducing the training matrix size, while utmost maintaining as good classification performance as that obtained when considering the original size. Regarding PS algorithm designing, technical literature reports at least three main methods (namely, compensation-based, edition-based, and hybrid) [27]. As have been mentioned throughout this paper, the whole process is carried out in such manner that the prototype selection results (reduced data matrices) can be stored directly into every WSN node.

In this work, in order to account for an enough coverage, we have chosen three representative algorithms of each method, as follows:

- `Condensation`: Condensed Nearest Neighbor (CNN), Reduced Nearest Neighbor (RNN), and Selective Nearest Neighbor (SNN).
- `Edition`: Edited Nearest Neighbor (ENN), All-k Edited Nearest Neighbors (AENN), and Iterative Partitioning Filter (IPF).
- `Hybrid`: Decremental Reduction Optimization Procedures 2 (DROP 2), Decremental Reduction Optimization Procedures 3 (DROP3), and Iterative Noise Filter based on the Fusion of Classifiers (INFFC).

3.3.3. Classification Algorithms

Classification algorithms can learn based on different criteria, having each of them representative algorithms [27]. Herein, we consider four criteria and their respective representative algorithm, namely:

- Distance-based: K-Nearest Neighbors (KNN).
- Model-based: Support Vector Machine (SVM).
- Density-based: Bayesian classifier (BC).
- Heuristic: Decision Tree (DT).

Given that the four aforementioned criteria are essentially different, a comparison of individual performances is necessary to identify the one(s) best fitting the nature of data and classification task. As well, it is of crucial interest to measuring the computational cost that each algorithm involves to be further implemented within the WSN node.

The database—obtained according to the pollution level—has been divided regarding the information acquired by the WSN nodes into 3 types (being our training labels): high, medium and low contamination. Therefore, if the system is located at different spots along the river, it can generate a map of the pollution status and estimate the river's course. Alternatively, if it is located statically, the system can determine, in hours, how the level of contamination varies with respect to the time of day.

4. Results and Discussion

In order to evaluate the behavior of each stage, we firstly discuss the data reduction in the training matrix. Subsequently, we show the outcome of our proposed analysis scheme, namely, the performance analysis using our defined metric (QMB) for determining the ideal algorithms for its implementation in the WSN nodes. Finally, we present the results of the final implementation of the system and the tests in real environments.

4.1. Data Reduction

The sensors were acquiring data during the months of July, August and September on random days. As a result, we obtained the data matrix called $\mathbf{Y} \in \mathbb{R}^{m \times n}$, where m is the number of instances, and n the number of measured variables (sensors). While, $\mathbf{L} \in \mathbb{R}^{m \times 1}$ is the tag vector. Thus, we have that $m = 507$, and $n = 4$. With these data, we implemented the PS algorithms in order to reduce the training matrix and processing time. In addition, to validate the classification criteria, we retained 20% of the $\mathbf{Y}$ matrix for performance testing. In succession, the matrix for the data scheme is $\mathbf{X} \in \mathbb{R}^{p \times n}$, where $p = 405$. Table 2 shows the summary of the PS algorithms results and find a new reduced data matrix $\mathbf{Z}$.

Accordingly, we have selected the CNN, DROP1 and DROP3 algorithms as they reach the highest percentages of reduction in the database. Figure 5 shows scatter plots of the initial data set and the reduced versions generated by CNN, DROP1 and DROP3.

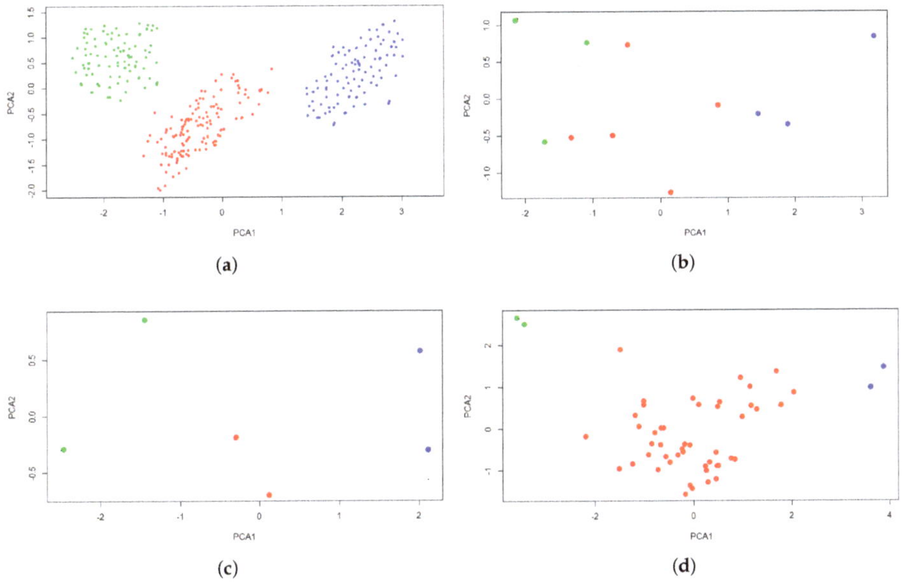

Figure 5. 2D scatter plots of resulting data matrices $\mathbf{Z}$ of the chosen prototype selection algorithms. (**a**) Data matrix X, (**b**) CNN, (**c**) DROP1, (**d**) DROP3.

Table 2. Analysis of PS algorithms in relation to optimization embedded computational resources.

PS Algorithm	Exec. Time (s)	Remv. Inst	% of Remv. Inst
AENN	3.17	0	0
BBNR	125.23	102	25.18
CNN	2.28	394	97.28
DROP1	130.63	399	98.51
DROP2	230.28	354	87.407
DROP3	264.97	354	87.40
ENG	250	210	51.85
ENN	0.72	0	0
RNN	2.39	394	97.28

4.2. Classification Performance

With the reduced data sets, we compared the classification performance using the aforementioned algorithms. Table 3 summarizes the results of the classifiers with cross-validation with ten random folds.

Table 3. Classifier's metrics.

Classifier	Matrix X%	CNN%	DROP1%	DROP3%
Accuracy				
k-NN	97.6	90.6	93.6	95
Bayesian classifier	95	87.5	82.6	0.99
Decision Trees	99.3	66.9	33.33	33.33
SVM (Polynomial kernel)	100	75	75.3	92.14
SVM (Sigmoide kernel)	100	75	92	100
Sensitivity				
k-NN	96.6	88.3	91.6	93.3
Bayesian classifier	93.3	75	76	97.3
Decision Trees	99.3	33	33.33	33.33
SVM (Polynomial kernel)	100	94	66.9	92.14
SVM (Sigmoide kernel)	100	50	90	100
Specificity				
k-NN	98.2	93.6	95.3	96.6
Bayesian classifier	96.6	88.6	88	99.3
Decision Trees	99.6	66.9	33.33	33.33
SVM (Polynomial kernel)	100	97	89	100
SVM (Sigmoide kernel)	100	100	94	100
Precision				
k-NN	96.3.0	98.3	89.3	93.3
Bayesian classifier	93.3	100	66.6	98.3
Decision Trees	93.3	33.9	33.33	33.33
SVM (Polynomial kernel)	100	93	89	93.13
SVM (Sigmoid kernel)	100	50	86.6	100

To graphically appreciate the results of the whole data processing scheme, just as done in previous works [11,14], we use the principal component analysis conventional algorithm as a dimensionality reduction approach to represent the original data over a lower-dimensional domain. Figure 6 presents scatter plots regarding the two first principal components to depict the decision borders generated by every considered classifier. This process is carried out for demonstration purposes in order to know the algorithms' ability to differentiate each label in an understandable way for the human being perception (visual-type in this case).

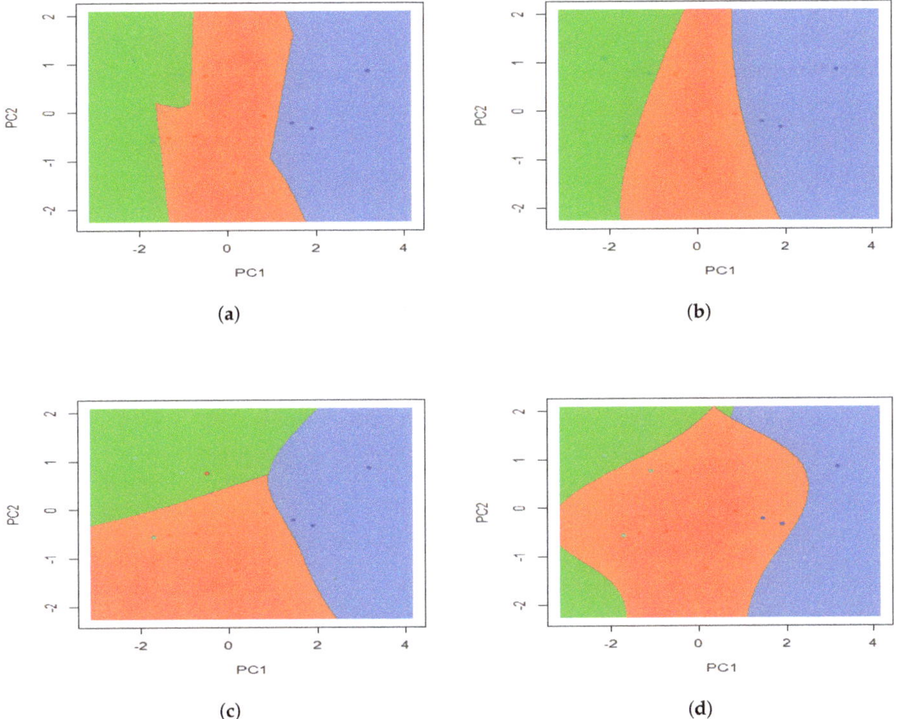

Figure 6. Decision borders for each classifier. Original data are embedded into a bi-dimensional space using PCA to graphically depict the classification ability of the considered algorithms. (**a**) k-NN, (**b**) Bayesian classifier, (**c**) SVM (Sigmoid kernel), (**d**) SVM (Polynomial kernel).

Numerical results of the joint performance of the prototype selection and data classification are summarized in Table 4.

Table 4. QMB analysis for every classifier along with the previously identified PS algorithms.

Classification	Exec. Time (s)	QMB Value	
algorithm		CNN%	DROP1%
k-NN	1.21	72.85	76.20
Bayesian	1.85	46.01	43.97
Decision Tree	0.77	42.06	42.63
SVM (polynomial kernel)	5.2	14.03	14.2
SVM (sigmoide kernel)	6.1	11.96	12.11

Discussion on performance measures: As can be seen in the Table 3, VSM reaches the best classification performance based on the considered metrics (100%). Nonetheless, its algorithm involves mathematical functions (known as kernel functions), which are not able to readily processed in a WSN. In this connection, the proposed QBM allows for warning about this computational cost in relation to the amount of data used to train the classification algorithm and the system response time when assigning the corresponding label to a new data from the sensors. This can be appreciated from the fact that by reducing the training matrix its performance decreases significantly. The same occurs for all the considered algorithms excepting for k-NN, whose distance-based nature is non-expensive in terms of computational cost. Furthermore, by using a reduced data matrix, k-NN considerably maintains its performance. Furthermore, it is clearly noted that DROP1 is the best-suited algorithm for prototype selection although its computational cost is very high. Hence, given the design settings and the embedded systems conditions, CNN is preferred and therefore selected as the algorithm for prototype selection, while k-NN is considered as the selected classification algorithm reaching a performance of 90.6% and a QBM value of 72.85%.

4.3. Implementation and Testing

Figure 7 depicts the functional architecture of the nodes using the proper, selected prototype selection algorithms, which are to be compiled within thereof. As can be appreciated, each node holds the data-acquisition sensor set. The data analysis and processing is as follows: The raw data is first filtered by using the Moving average filter, which, in this case, is enough to remove the components (artifacts) related to reading errors and noise. Subsequently, data are classified by the algorithm k-NN, which assigns a label and decides about the predicted level of water contamination according to the training database and following a distance-based, majority-vote-driven approach. Then, data undergo an additional processing via CNN to determine whether the training database can be improved by removing instances exhibiting negligible relevance regarding either the subsequent classification task or the intrinsic knowledge they may hold. Finally, the output information is converted into a character string together with its label to be sent by the GSM network to the external server and display the data obtained from each sensor and the decision made. It is worth highlighting that the node to be monitored can be selected through the interface.

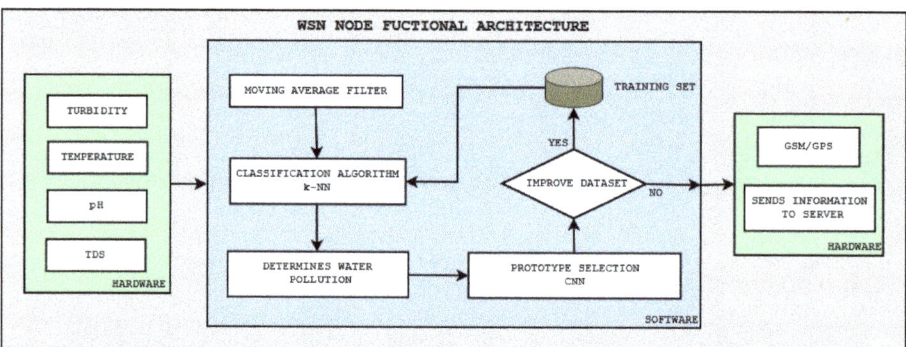

Figure 7. WSN node functional architecture incorporating the workflow of the in-situ data analysis and processing and mainly consisting in filtering, prototype selection and classification.

In the overall work-flow of our approach, the need for using an external sever lies in the fact that optimizing resource consumption at the in-situ analysis (directly on WSN Nodes) entails performing offline data processing tasks, mainly, at three specific points. The first one is when collecting data from each WSN node, being its main function the storing of such information (which—at this extent—corresponds to the outcomes of reading-errors-filtering stage produced by the moving average filter). The second one is the offline, exhaustive running, and comparison of classification algorithms to identify the ones reaching a good compromise between accuracy and computational cost, and therefore, being adequate to be directly implemented into the WSN nodes. Finally, as the third point, the server is used for information visualization purposes (displaying numerically and graphically the acquired data, the decision (classification) made by each node and the river pollution historical). This information is also stored in the server. Of course, those algorithms identified as adequate ones at the second point are the ones that are finally incorporated into the WSN nodes.

Once performed the data analysis procedures, we integrate all sensors into a PCB board incorporating an `Arduino Uno` as a processor unit. A view of the developed WSN node can be seen in Figure 8.

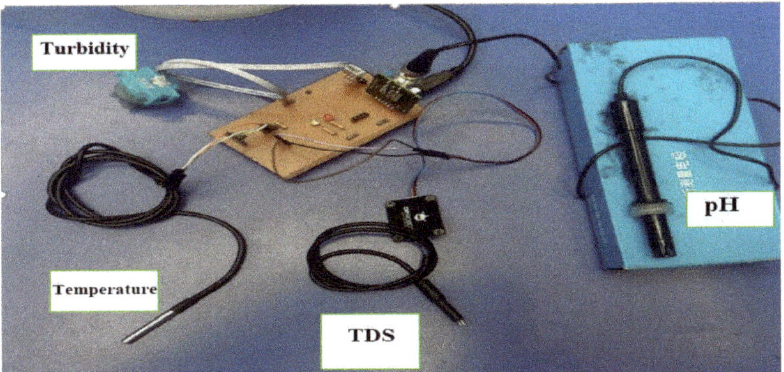

Figure 8. View of the WSN node including the four considered sensors and the processor.

The developed WSN has a considerably high operating consumption for a LiPo-type battery. To increase the life time of both the system and the battery, energy saving modes are used inside the Arduino board that handles the sensor activation. To enable such modes, we consider the use of timers, which work as an internal clock determining the data-acquisition-and-sending timing, and therefore limit current consumption. Hence the power consumption of every single sensor and the processor should be considered. In normal operation conditions, the total electric current consumption (considering all the sensors) amounts to 110 mA, while the GPS-GSM module and the Arduino require 40 mA and 45 mA, respectively. Meanwhile, when the battery saving system is enabled, the sensors and the GPS module are not is used, and thus only the Arduino works and is fed with 15mA. As stated in [28], the following equation relates the battery life time with the total power consumption (P):

$$P = \frac{(T_{on} * I_{on}) + (T_{sleep} * I_{sleep})}{T_{on} + T_{sleep}}, \tag{7}$$

where T_{on}, T_{sleep}, I_{on}, and I_{sleep} stand respectively for Normal Consumption Time, Sleep Consumption Time, Current Consumption at Normal Conditions, and Current Intensity Sleeping Consumption.

As explained in Section 3.3, the system is on during 10 min and then remains in battery saving mode. As a result, the system consumes 78.45 mA per hour. If the used battery is 5 volts at 1000 mA, the system can work continuously for 12.73 h. However, the system is activated only four times per day (early morning, mid-morning, afternoon and night), that is, it only works for 4 h a day. As a result, the system can remain for at least 3 days with no requiring battery manager support. As an advantageous aspect of our system we may say that, when implemented with a solar panel powering the battery, there is experimental evidence that it can work up to 4 months with no discharging or critical battery issues.

Subsequently, over the implemented system, we store the training dataset obtained after running the CNN algorithm, which is to denoted $\mathbf{Z} \in \mathbb{R}^{s \times n}$, by setting the number of prototypes as $s = 11$. At this extent, CNN algorithm is considered as an recommendable approach, since its execution time is the least while its ability to reduce the dataset instances is proper enough. Consequently, if the system requires to be reconfigured to train the classification algorithm model, the CNN algorithm can be compiled readily on the WSN network with no entailing extra battery consumption or diminishing the system performance. Then, we implemented the Bayesian classifier so that it can make system decisions concerning the tag assigned by location. Thus, we can determine the contamination levels (high, medium, low) using the nodes along the river. Since the system is intended to be waterproof, we use a river buoy to keep the system afloat. At its upper part, we install the solar panel and the GPS-GSM communication antenna. Furthermore, the nodes are anchored using an ironwork attached to the river stones, as shown in Figure 9.

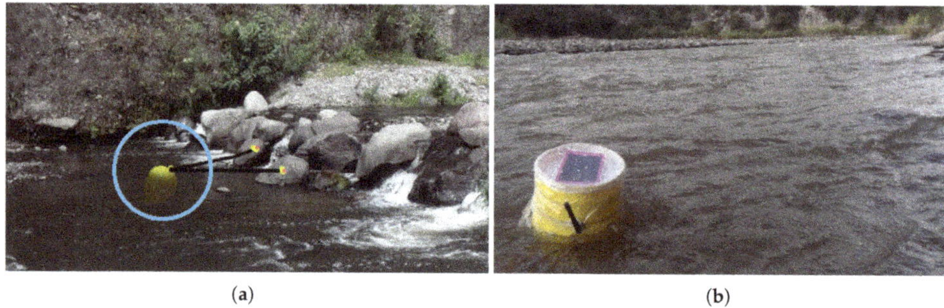

(a)　　　　　　　　　　　　　　　　　(b)

Figure 9. Anchored node acquiring and sending data to interface. (**a**) Simulation. (**b**) Real conditions.

Besides, for displaying purposes, we develop a monitoring interface in Processing using a local server that downloads and visualizes the information from the server. In this interface, we show the status of each sensor, the node location, and the level of contamination of the river. Figure 10 summarizes both the sensor testing and the visual interface with the decision taken.

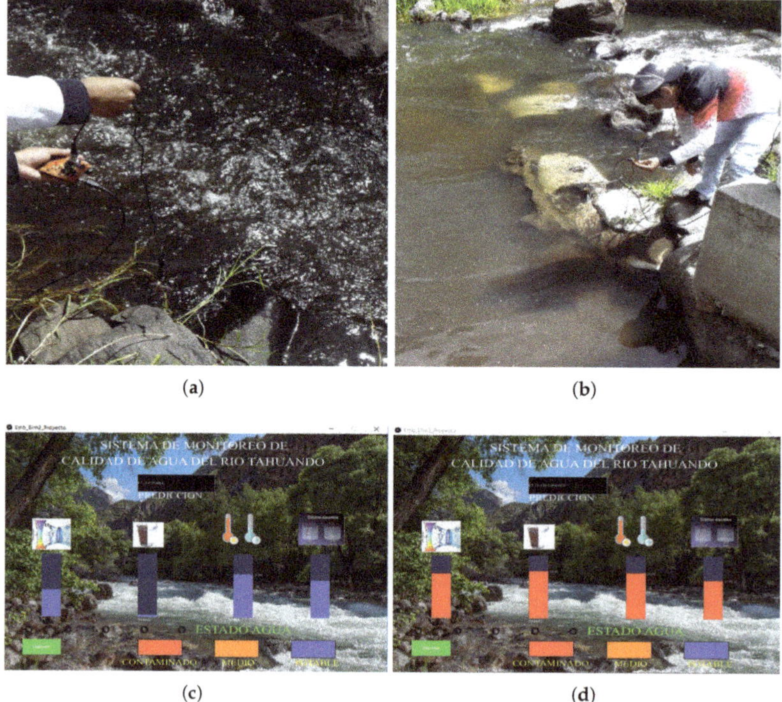

(a)　　　　　　　　　　　　　　　　　(b)

(c)　　　　　　　　　　　　　　　　　(d)

Figure 10. System testing and visual interface. (**a**) Testing embedded system developed in the rural sector. (**b**) Testing embedded system developed in urban sector. (**c**) Visual interface showing low level of contamination. (**d**) Visual interface showing high level of contamination.

For a more extensive analysis, we move the nodes throughout the river to assign a color label, based on the contamination level, as follows: red refers to high contamination, yellow to medium, and green to null or low pollution. Accordingly, Figure 11 shoes the contamination levels along the case-study river. As a relevant result, we identify that at the Campiña church zone there is already a high level of pollution.

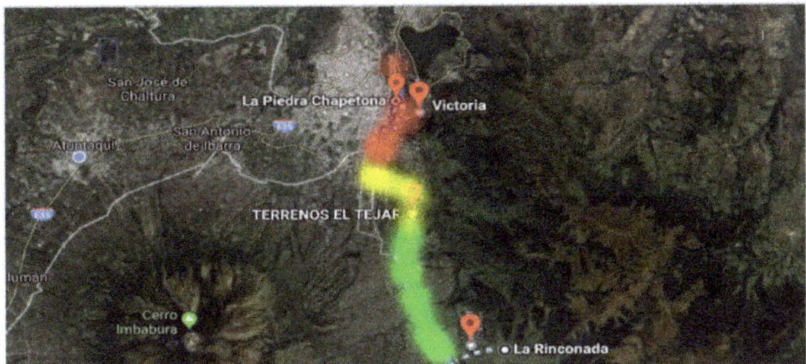

Figure 11. Tahuando river conditions along its stream bed.

Finally, with all nodes running, we daily capture data to observe the maximum values, in order to detect the hours of the day with highest contamination, which are in line with the human's work schedules. Figure 12 shows the pH, Temperature, and NTU values registered by the sensors during a whole day.

It is worth mentioning that our system may exhibit failures regarding the loss of signal from the GPS-GSM module when restarting it to carry out the data acquisition. To overcome this drawback, we follow a heuristic sensor calibration procedure as follows: On one hand, when activated, the system first turns on the GPS-GSM module so that there would be enough time to re-link to the GSM network and send back a status indicator signal. On the other hand, the length of the cables connected to the sensors was initially very long. This caused that when the volume of water decreased, cables descended to the bottom of the river and got brushed against stones. Consequently, since the length of the system-incorporated sensor is between 2 and 5 cm, an excessive wear on the sensors is induced. To cope with this issue, we search for and identify points where the river depth is the least possible varying, and is not prone to water stagnation.

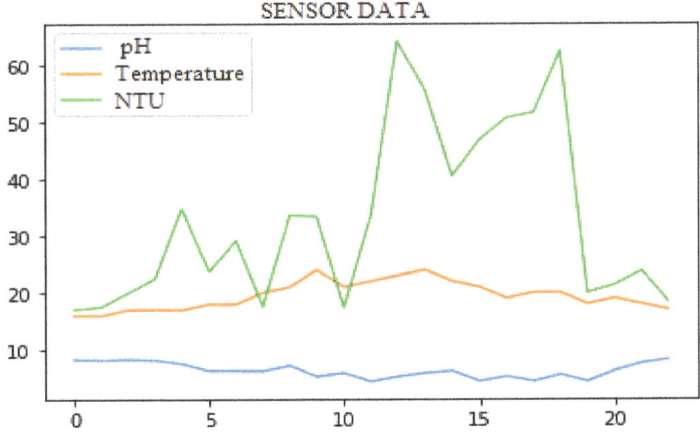

Figure 12. Sensor-generated data acquired per hour during a day.

5. Final Remarks

In this work, we present the complete design and validation of an intelligent wireless sensor network (WSN) system to measure the contamination levels of a river. Particularly, the Tahuando River is of interest. Broadly speaking, the proposed system involves two stages: electronic device implementation, and data analysis.

For the electronic design, since the case-study river may have high levels of pollution, as well as it may occur significant variations depending on the hours of the day, and zones of its route, we implement several WSN nodes for acquiring the river's conditions information by covering a meaningful zone and within a wide enough range of time. In this sense, we both calibrate and tune the sensors for a correct data collection. Additionally, we experimentally demonstrate that our data reading schedules were adequate for detecting higher pollution hours. Furthermore, we highlight that the river buoys is a key element to meet the node's permeability requirements as well as to enable the proper functioning of each WSN node.

Regarding the proposed data analysis scheme, we demonstrate that a classifier together with a prototype selection is suitable for a WSN-based water-quality monitoring system. It is reached a good trade-off between the computational resource usage (as the training matrix size is reduced to meet the system operation conditions), and the classification performance at detecting the pollution levels along the river. In addition, given the network coverage, the proposed system is able to send information from the WSN node to the server. Therefore, the filtered data can be visualized in an interface, and an in-situ analysis becomes possible. It is important to mention that the server is only for data visualization purposes and does not have the implementation of machine learning algorithms.

As a future work, the battery life is to be more carefully considered by exploring both different methods of extending its duration and alternatives sources of energy to supply the nodes (i.e., using the water flow to generate energy). A large number of nodes and wider coverage (located at different water resources around the province of Imbabura, Ecuador) is highly desirable for further In addition, we are intended to a seek for alternatives to mitigate system affectations due to disturbances caused by the presence of unexpected individuals (either people or animals), as so far our readily solution has been to locating the system in a hardly visible and difficult-to-access spot.

Author Contributions: P.D.R.-M.: Conceptualization, methodology, software, formal analysis, investigation, writing—original draft preparation, visualization, resources, V.F.L.-B.: investigation, supervision, project administration, J.A.R.: formal analysis, writing—review, visualization, D.H.P.-O.: validation, writing—review, methodology supervision, project administration. All authors have read and agreed to the published version of the manuscript.

Funding: This research received no external funding.

Acknowledgments: This work is supported by the Smart Data Analysis Systems Group—SDAS Research Group (http://sdas-group.com).

Conflicts of Interest: The authors declare no conflict of interest.

References

1. Venancio Cruz, D.; Rivelino Gomes de Oliveira, M.; Cunha Filho, M.; Venancio da Cruz, D. Monitoring pH with quality control based on Geostatistics Methodology. *IEEE Lat. Am. Trans.* **2016**, *14*, 4787–4791. [CrossRef]

2. Yang, C.; Wang, X. The water quality and pollution character in QingShuiHai lake valley-typical urban drinking water sources. In Proceedings of the 2011 International Conference on Remote Sensing, Environment and Transportation Engineering, Nanjing, China, 24–26 June 2011; pp. 7287–7291.

3. Zhang, Z.; Zhang, F.; Xu, C.; Xu, J.; Zhang, W.; Qi, Q. Study on the water environment capacity for the typical watershed in Taizihe River. In Proceedings of the 2011 International Symposium on Water Resource and Environmental Protection, Xi'an, China, 20–22 May 2011; Volume 1, pp. 486–488.

4.   Randhawa, S.; Sandha, S.S.; Srivastava, B. A Multi-sensor Process for In-Situ Monitoring of Water Pollution in Rivers or Lakes for High-Resolution Quantitative and Qualitative Water Quality Data. In Proceedings of the 2016 IEEE Intl Conference on Computational Science and Engineering (CSE) and IEEE Intl Conference on Embedded and Ubiquitous Computing (EUC) and 15th Intl Symposium on Distributed Computing and Applications for Business Engineering (DCABES), Paris, France, 24–26 August 2016; pp. 122–129. [CrossRef]

5.   Zhai, C.; Huang, Q.; Chang, J.; Gao, F. The study of water resources reasonable allocation of BaoJi area in Wei River with considering the ecology base flow. In Proceedings of the 2011 International Symposium on Water Resource and Environmental Protection, Xi'an, China, 20–22 May 2011; pp. 816–818. [CrossRef]

6.   Guo, W.; Chen, J.; Sheng, Y.; Wang, J. Integrated evaluation of water quality and quantity of the Wei River reach in Shaanxi Province. In Proceedings of the 2011 International Symposium on Water Resource and Environmental Protection, Xi'an, China, 20–22 May 2011; pp. 863–866. [CrossRef]

7.   Zhang, H.; Xie, X.; Hou, J. Water pollution accident control and urban safety water supply. In Proceedings of the 2011 2nd IEEE International Conference on Emergency Management and Management Sciences, Beijing, China, 8–10 August 2011; pp. 37–40.

8.   De Agua, S. Biblioteca—Secretaría del Agua. Available online: https://www.agua.gob.ec/ (accessed on 1 January 2020).

9.   Wang, J.; Guo, X.; Zhao, W.; Meng, X. Research on water environmental quality evaluation and characteristics analysis of TongHui River. In Proceedings of the 2011 International Symposium on Water Resource and Environmental Protection, Xi'an, China, 20–22 May 2011; pp. 1066–1069. [CrossRef]

10. Taufiqurrahman; Tamami, N.; Putra, D.A.; Harsono, T. Smart sensor device for detection of water quality as anticipation of disaster environment pollution. In Proceedings of the 2016 International Electronics Symposium (IES), Denpasar, Indonesia, 29–30 September 2016; pp. 87–92. [CrossRef]

11. Rosero-Montalvo, P.D.; Pijal-Rojas, J.; Vasquez-Ayala, C.; Maya, E.; Pupiales, C.; Suarez, L.; Benitez-Pereira, H.; Peluffo-Ordonez, D. Wireless Sensor Networks for Irrigation in Crops Using Multivariate Regression Models. In Proceedings of the 2018 IEEE Third Ecuador Technical Chapters Meeting (ETCM), Cuenca, Ecuador, 15–19 October 2018; pp. 1–6. [CrossRef]

12. Ragnoli, M.; Barile, G.; Leoni, A.; Ferri, G.; Stornelli, V. An Autonomous Low-Power LoRa-Based Flood-Monitoring System. *Low Power* **2020**, *10*, 15. [CrossRef]

13. Alippi, C. *Intelligence for Embedded Systems*; Springer: Berlin/Heidelberg, Germany, 2014; pp. 1–283. [CrossRef]

14. Rosero-Montalvo, P.D.; Batista, V.F.L.; Rosero, E.A.; Jaramillo, E.D.; Caraguay, J.A.; Pijal-Rojas, J.; Peluffo-Ordóñez, D.H. *Intelligence in Embedded Systems: Overview and Applications*; Springer: Cham, Switzerland, 2019; pp. 874–883. [CrossRef]

15. Guo, M.; Zhou, X. Research on the water environment capacity of Chanba River downstream. In Proceedings of the 2011 International Conference on Electric Technology and Civil Engineering (ICETCE), Lushan, China, 22–24 April 2011; pp. 4411–4414. [CrossRef]

16. Patel, H.J.; Dabhi, V.K.; Prajapati, H.B. River Water Pollution Analysis using High Resolution Satellite Images : A Survey. In Proceedings of the 2019 5th International Conference on Advanced Computing & Communication Systems (ICACCS), Coimbatore, India, 15–16 March 2019; pp. 520–525. [CrossRef]

17. Shukla, A.K.; Ojha, C.S.P.; Garg, R.D. Surface water quality assessment of Ganga River Basin, India using index mapping. In Proceedings of the 2017 IEEE International Geoscience and Remote Sensing Symposium (IGARSS), Fort Worth, TX, USA, 23–28 July 2017; pp. 5609–5612. [CrossRef]

18. Lin, Z.; Wang, W.; Yin, H.; Jiang, S.; Jiao, G.; Yu, J. Design of Monitoring System for Rural Drinking Water Source Based on WSN. In Proceedings of the 2017 International Conference on Computer Network, Electronic and Automation (ICCNEA), Xi'an, China, 23–25 September 2017; pp. 289–293.

19. Sowmya, C.; Naidu, C.D.; Somineni, R.P.; Reddy, D.R. Implementation of Wireless Sensor Network for Real Time Overhead Tank Water Quality Monitoring. In Proceedings of the 2017 IEEE 7th International Advance Computing Conference (IACC), Hyderabad, India, 5–7 January 2017; pp. 546–551.

20. Chen, F.; Wen, F.; Jia, H. Algorithm of Data Compression Based on Multiple Principal Component Analysis over the WSN. In Proceedings of the 2010 6th International Conference on Wireless Communications Networking and Mobile Computing (WiCOM), Chengdu, China, 14 October 2010; pp. 1–4.

21. Kadir, E.A.; Irie, H.; Rosa, S.L. River Water Pollution Monitoring using Multiple Sensor System of WSNs (Case: Siak River, Indonesia). In Proceedings of the 2019 6th International Conference on Electrical Engineering, Computer Science and Informatics (EECSI), Bandung, Indonesia, 18–20 September 2019; pp. 75–79.

22. Zhang, Z. Data Fusion Optimization Analysis of Wireless Sensor Networks Based on Joint DS Evidence Theory and Matrix Analysis. In Proceedings of the 2019 4th International Conference on Mechanical, Control and Computer Engineering (ICMCCE), Hohhot, China, 24–26 October 2019; pp. 689–6894.

23. Torres, A.J.; Quezada, M.; Carrion, L.; Coronel, I.; Barragen, A. AHP analysis to minimize the effects produced by the textile industry in the rivers of Cuenca city. In Proceedings of the 2017 IEEE Mexican Humanitarian Technology Conference (MHTC), Puebla, Mexico, 29–31 March 2017; pp. 94–101. [CrossRef]

24. De Estadística y sensos, I.N. Fasículo Provincial de Imbabura. Available online: https://www.ecuadorencifras.gob.ec/institucional/home/ (accessed on 1 January 2020).

25. Encarnación, D.; Enríquez, J.; Suarez, L. *Derecho De La Naturaleza: Caso Rio Tahuando*; Technical Report; Universidad Andina Simń Bolivar Ambato, Ecuador, 2012.

26. Liu, J.; Deng, Z. Self-tuning weighted measurement fusion Wiener filter for autoregressive moving average signals with coloured noise and its convergence analysis. *IET Control. Theory Appl.* **2012**, *6*, 1899–1908. [CrossRef]

27. Rosero-Montalvo, P.D.; López-Batista, V.F.; Peluffo-Ordóñez, D.H.; Erazo-Chamorro, V.C.; Arciniega-Rocha, R.P. Multivariate Approach to Alcohol Detection in Drivers by Sensors and Artificial Vision. In *From Bioinspired Systems and Biomedical Applications to Machine Learning*; Ferrández Vicente, J.M., Álvarez-Sánchez, J.R., de la Paz López, F., Toledo Moreo, J., Adeli, H., Eds.; Springer International Publishing: Cham, Switzerland, 2019; pp. 234–243.

28. Antolín, D.; Medrano, N.; Calvo, B. Analysis of the operating life for battery-operated wireless sensor nodes. In Proceedings of the IECON 2013—39th Annual Conference of the IEEE Industrial Electronics Society, Vienna, Austria, 10–13 November 2013; pp. 3883–3886.

remote sensing

Article

Groundwater Potential Mapping Using Remote Sensing and GIS-Based Machine Learning Techniques

Sunmin Lee [1,2], Yunjung Hyun [3], Saro Lee [4,5] and Moung-Jin Lee [6,*]

1   Department of Geoinformatics, University of Seoul, 163 Seoulsiripdaero, Dongdaemun-gu, Seoul 02504, Korea; smilee@uos.ac.kr
2   Center for Environmental Assessment Monitoring, Environmental Assessment Group, Korea Environment Institute (KEI), 370 Sicheong-daero, Sejong-si 30147, Korea
3   Department of Land and Water Environment Research, Korea Environment Institute (KEI), 370 Sicheong-daero, Sejong-si 30147, Korea; yjhyun@kei.re.kr
4   Geoscience Platform Division, Korea Institute of Geoscience and Mineral Resources (KIGAM), 124, Gwahak-ro Yuseong-gu, Daejeon 34132, Korea; leesaro@kigam.re.kr
5   Department of Geophysical Exploration, Korea University of Science and Technology, 217 Gajeong-ro Yuseong-gu, Daejeon 34113, Korea
6   Center for Environmental Data Strategy, Korea Environment Institute (KEI), 370 Sicheong-daero, Sejong-si 30147, Korea
*   Correspondence: leemj@kei.re.kr; Tel.: +82-44-415-7314

Received: 17 February 2020; Accepted: 7 April 2020; Published: 8 April 2020

Abstract: Adequate groundwater development for the rural population is essential because groundwater is an important source of drinking water and agricultural water. In this study, ensemble models of decision tree-based machine learning algorithms were used with geographic information system (GIS) to map and test groundwater yield potential in Yangpyeong-gun, South Korea. Groundwater control factors derived from remote sensing data were used for mapping, including nine topographic factors, two hydrological factors, forest type, soil material, land use, and two geological factors. A total of 53 well locations with both specific capacity (SPC) data and transmissivity (T) data were selected and randomly divided into two classes for model training (70%) and testing (30%). First, the frequency ratio (FR) was calculated for SPC and T, and then the boosted classification tree (BCT) method of the machine learning model was applied. In addition, an ensemble model, FR-BCT, was applied to generate and compare groundwater potential maps. Model performance was evaluated using the receiver operating characteristic (ROC) method. To test the model, the area under the ROC curve was calculated; the curve for the predicted dataset of SPC showed values of 80.48% and 87.75% for the BCT and FR-BCT models, respectively. The accuracy rates from T were 72.27% and 81.49% for the BCT and FR-BCT models, respectively. Both the BCT and FR-BCT models measured the contributions of individual groundwater control factors, which showed that soil was the most influential factor. The machine learning techniques used in this study showed effective modeling of groundwater potential in areas where data are relatively scarce. The results of this study may be used for sustainable development of groundwater resources by identifying areas of high groundwater potential.

Keywords: groundwater potential; specific capacity; machine learning; boosted tree; ensemble models

1. Introduction

Because groundwater has less exposure to pollution than surface water, it is considered a valuable natural resource for agriculture in many communities [1]. Especially during the drought season, a continuous supply of groundwater is important in agricultural areas. The study area in this investigation, Gyeonggi-do, has recently suffered from damage to agricultural land due to increasing drought. In 2018, widespread damage to crops due to heat waves and drought continued throughout the year, and the average storage rate in 339 reservoirs in Gyeonggi-do was 59% of capacity, which was only 76% of the normal level [2].

Groundwater is a good water resource because it can stably supply the required amount of high-quality water; thus, appropriate water conservation plans are essential for the sustainable use of groundwater [3]. In many areas, the main causes of groundwater depletion are excessive groundwater extraction and unsuitable aquifer recharge [4]. Therefore, accurate estimation and prediction of groundwater recharge should be carried out to support efficient use and systematic management of groundwater resources. From this perspective, groundwater potential mapping using yield data is important. Yield data include extraction volume and the velocity of groundwater at various measurement points. Groundwater yield depends on geological, topographic, and anthropogenic factors specific to the area, and is also related to groundwater potential [5].

In practical terms, groundwater is less accessible than surface water. Groundwater can be presumed by detecting gravity anomalies such as Gravity Recovery and Climate Experiment (GRACE) [6–8]; however, a local groundwater potential map is essential for regional management of groundwater. Thus, studies on the distribution and prediction of groundwater resources have been limited to local scales based on data obtained from point measurements (e.g., meteorological stations, flow measurement points, and groundwater level monitors) [9,10]. In recent years, areal distribution analysis data obtained through remote sensing have been used for global prediction of the water resource distribution in combination with various machine learning techniques, albeit with high uncertainty. To overcome the limitation of groundwater resource surveys based on local information, these data can be converted into global distribution data using satellite imagery. Remote sensing generally produces data in the form of grids or regions, which can be converted into distribution patterns through various processing methods such as machine learning algorithms. By applying the characteristics of remote sensing data to groundwater resources, point-based groundwater hydrological modeling can be extended to the global scale. Therefore, using existing groundwater yield data, it is possible to make regional and local predictions with remote sensing-based methods.

For groundwater potential mapping, a variety of techniques have been applied, including direct drilling for hydrological testing and geophysical models [11,12]. Such methods are suitable for identifying the hydrological characteristics of groundwater, but have high costs in time and money [13,14]. In recent years, studies related to groundwater potential have been conducted using machine learning models with available historical data on groundwater wells with geographic information systems (GIS) [15,16]. GIS technologies have been used for quantitative analysis of spatial distributions in environmental, geological, and hydrological studies [17–19]. One limitation of data-based analysis of groundwater is insufficient availability of data for analysis [20]; groundwater yield varies with hydrological conditions and recharge sources, which have been measured in a limited number of groundwater wells [21]. Therefore, using various models to predict groundwater yield accurately and identifying the optimal model for water resource evaluation in a given region are essential to effective water resource management.

For this reason, studies related to groundwater potential mapping with various data models have become increasingly common [22–24]. Numerous factors that affect groundwater potential have been proposed based on various data modeling methodologies, including statistical models, probabilistic models, machine learning models, and data mining models; yield and spring or well location data are also widely used as groundwater potential indicators. Due to the characteristics of remote sensing and groundwater, groundwater could be indirectly monitored by using remote sensing; much research has been conducted through thematic maps related to groundwater based on remote sensing data and groundwater potential was estimated by reducing the uncertainties [25–27].

The frequency ratio (FR) model is a representative statistical model applied to groundwater potential mapping [26,28,29]. The relationship between groundwater conditioning factors and groundwater potential could be analyzed using basic statistical and probabilistic models, including FR, weight of evidence [30], evidential belief function [31], and logistic regression [32] models. Furthermore, the recent exponential increase in available data has led to identification of data types and data processing techniques that can support decision-making. Several studies in this area have applied machine learning methods such as machine learning models, while artificial neural networks [33] and support vector machines [34] have been widely applied to groundwater potential mapping. Some studies have also used analytical hierarchy methods, which are expertise-based methods requiring a deep understanding of the study area [35,36]. Recently, hybrid and ensemble models that combine or develop existing methodologies have been applied for groundwater potential mapping [37–39]. This paper also uses a hybrid methodology in this respect.

When performing groundwater potential mapping through modeling, the results show poor generalizability without proper training samples. In such cases, the accuracy for training data is high but the testing results show significantly lower accuracy. To overcome the lack of data, robust models built upon basic models have recently been developed and compared [40]. Typically, an ensemble model using learner sequences is developed; voting, bagging, and adaptive boosting are representative ensemble methods that can be applied to various base learners [41]. In this way, unlabeled cases are identified via self-learning by combining information from labeled cases so that the labeled training set is magnified in each iteration until the entire dataset is labeled. This method, which was applied in the present study, could be effective for data-scarce areas because it allows modeling using less data than other approaches.

Previous studies conducted on groundwater recharge and yield have used enough field survey data targeted at adjacent areas. However, these studies are subordinate to field surveys and are not intended to reduce spatial uncertainty on groundwater. Therefore, the purpose of this study was to map and test groundwater yield potential in Yangpyeong-gun, South Korea, using spatial data analysis in a GIS environment. This study processed and analyzed officially published groundwater yield data using remote sensing and GIS to reduce the uncertainty of the data itself. In addition, one of the latest machine learning models, boosted tree method, was applied to predict large areas of low uncertainty using pumping test data from 53 wells; groundwater yield potential is the major issue of this study. The results of this study could provide a scientific basis for efficient use and systematic management of groundwater resources.

2. Study Area

South Korea consists of eight administrative districts, labeled '-do', which are made up of local administrative districts, labeled with '-si', '-gun', and '-gu'. The study area, Yangpyeong-gun, is located about 50 km from Seoul, in the northeastern part of Gyeonggi-do (Figure 1). Yangpyeong-gun is surrounded by Hongcheon-gun in Gangwon-do to the northeast, Hoengseong-gun in Gangwon-do to the east, Wonju-si in Gangwon-do to the southeast, and Gapyeong-gun to the north. Yangpyeong-gun contains rugged mountainous areas such as Yongmunsan (1157 m), Bongmyun (856 m), and Baekunbong (940 m), and the Namhan River flows from the south to the northwest of the district. About 90% of the total area of Yangpyeong-gun is a green zone covering the protected headwater area of the Han River; this area has a well-preserved and clean natural environment due to legal and institutional regulations [42].

Yangpyeong-gun covers approximately 878 km^2, and the amount of groundwater used in this area is 41,503,946 m^3/year. The groundwater use per unit area is 47,258 m^3/km^2 annually and 129 m^3/km^2 daily [43]. Groundwater in Gyeonggi-do is used primarily for agricultural purposes in numerous agricultural areas, including Anseong-si, Yangpyeong-gun, Icheon-si, and Yeoju-si. Among all districts in South Korea, Yangpyeong-gun (10,725) has the second highest number of groundwater facilities for agricultural use after Anseong-si [43].

In Yangpyeong-gun, a preliminary survey of available groundwater resources was conducted from December 2017 to June 2018 for drought response and to prevent unplanned development. Among 35 districts prone to drought, 25 were selected based on the feasibility of surveying the target district and the response rate of residents. A resistivity survey (vertical and dipole survey) was conducted to select locations for large-scale groundwater storage.

In Yangpyeong-gun, Gyeonggi-do, Kyonggi massif metamorphic rocks of Precambrian age and an intrusive body of Mesozoic Triassic gabbro and syenite are found. Precambrian Kyonggi massif metamorphic rocks consist of the Paleozoic sequence of Yongmunsan and unconformity of Jang-Rak. The main constituent rocks are banded gneiss, migmatitic gneiss, augen gneiss, mica schist, and quartzite. These rocks underwent metamorphism in the Paleozoic and Mesozoic Triassic, when the landmasses of North China and South China collided.

Groundwater development requires continuous management for sustainable supply of water rather than short-term measures at the time of drought. Specifically, preliminary investigation is needed in drought-prone areas and areas of high importance for agricultural water usage in Gyeonggi-do. To mount an effective response to agricultural drought, a groundwater management plan that ensures sustainable use of agricultural groundwater prior to drought is needed [44]. In this study, continuous groundwater potential data in the study area were used as primary data for a groundwater abundance survey, and could further be used to establish a groundwater development plan.

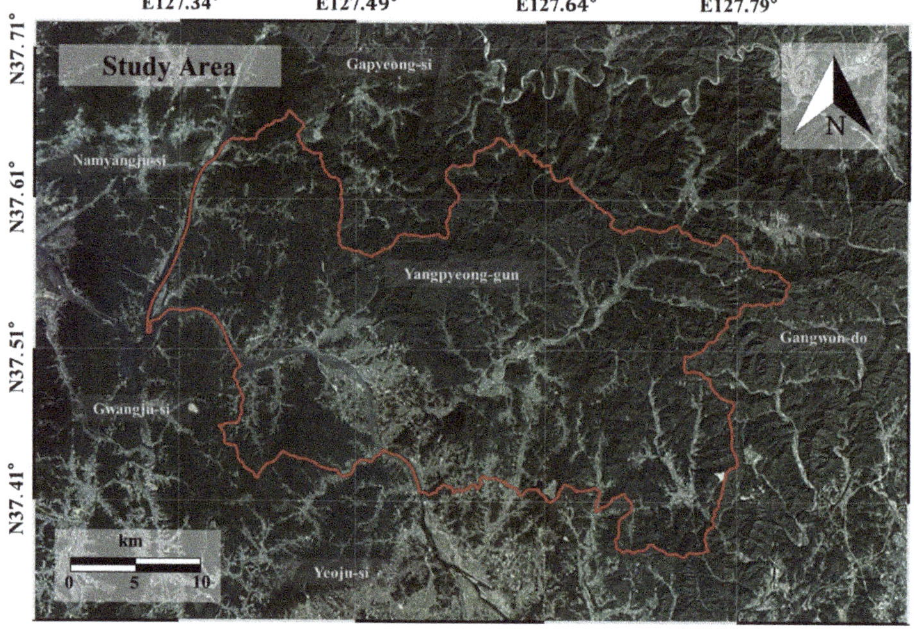

Figure 1. Study area.

3. Data

3.1. Groundwater Potential Analysis Based on Remote Sensing Data

Various thematic maps constructed using remote sensing source data were applied to machine learning techniques in this study. Recently, high-resolution aerial photographs were used to produce thematic maps of spatial data. Topographic maps were produced through numerical mapping using aerial photographs taken in 2006, with corrections and supplemental data collected through field surveys. Forest and soil maps were also constructed using spatial data generated through field surveys along with aerial photography. For land use maps, aerial photographs taken in 2012 were classified using image classification techniques, and their quality was verified using additional high-resolution satellite images from KOMPSAT-2 and KOMPSAT-3 as well as digital topographic maps. Meanwhile, geological maps were produced from field surveys and historical records using base maps generated from aerial photographs. Groundwater yield is a measure of groundwater pumping capacity, which could be stored in aquifers. In this study, groundwater yield potential modeling using machine learning was performed with spatial data generated via remote sensing and GIS such as soil, land cover, and geological maps, as described above.

3.2. Groundwater Well Data from in Situ Sampling

Groundwater pumped from wells in the study area is used mainly for agricultural purposes and domestic drinking water. Groundwater well data were collected for specific capacity (SPC) (53 wells) and transmissivity (T) (53 wells) from the basic survey report of Yangpyeong-gun [45]. The main use of the groundwater in this area is agricultural, so groundwater surveys are conducted between spring and summer, and our data was obtained between June and August. In the training and testing subsets, yield values above 3.8 and 3.42 (30 m^3/h) above the median value were considered for yields based on the dependent variables of SPC and T, respectively, which are two different indexes measured in different ways. Groundwater pumping test data used in this study were generated and published from the national groundwater observation and survey data by local governments conducted by Korea Water Resources Corporation (K-water).

SPC data include geographic location coordinates of individual wells and groundwater yield derived from pumping tests. SPC often indicates well performance, because it refers to the amount of water that a well can produce per unit of drawdown. SPC is calculated by dividing the pumping discharge by the drawdown, in units of liters per minute (LPM) per meter, as follows:

$$\mathrm{SPC} = \frac{Q}{S} \tag{1}$$

where Q is discharge (unit: LPM) and S is drawdown (unit: m). A low SPC value indicates that more energy is required for pumping. During a drawdown test to determine SPC, pumping should be maintained at a constant speed for a certain period of time, at least 24 h, with little change in drawdown. SPC data acquired during the pumping test can be used to estimate T and identify potential aquifer issues.

T represents the flow rate under a unit hydraulic gradient through a unit width of aquifer of a certain thickness [46]. Hydraulic conductivity (K) is a measure of the water transmission capacity of an aquifer. T of an aquifer is equal to the hydraulic conductivity multiplied by the thickness of the aquifer.

$$K\prime(x,y) = \frac{1}{b} \int_0^b K(x,y,z)dz \tag{2}$$

$$T = Kb, \tag{3}$$

where T is transmissivity, K is hydraulic conductivity, and b is aquifer thickness. Less drawdown and a thicker aquifer lead to higher T values. It is possible to estimate the amount of water flowing through the unit thickness of the aquifer by combining Equation (3) with Darcy's law.

SPC and T data were separately applied to the FR, boosted tree (BT), and ensemble models in this study; both SPC and T are used in this study in order to consider various aspects of groundwater. The locations of groundwater wells in the study area are shown in Figure 2. Yield data were randomly divided into a training data subset (70%) and a testing data subset (30%), as is the usual division in machine learning methodologies [16,47]. In the training data subset, 37 wells each were represented in SPC and T data, respectively; 16 wells were used to test the models.

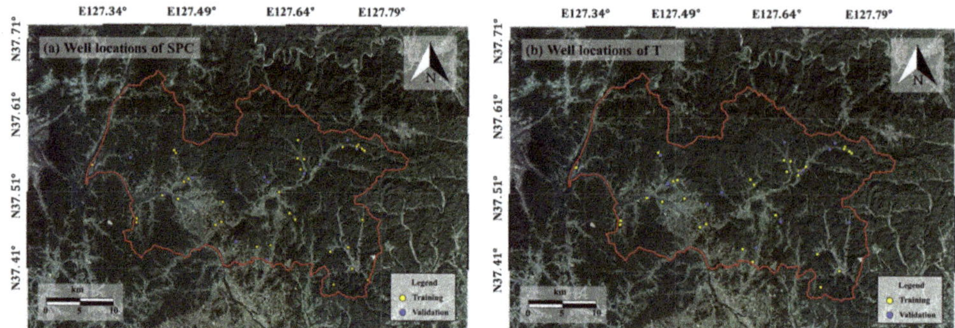

Figure 2. Locations of groundwater wells sampled for: (**a**) SPC and (**b**) T data.

3.3. Groundwater Conditioning Factors

Various groundwater conditioning factors were used for groundwater potential modeling in this study (Table 1). Topographical, geological, hydrological, and land cover factors are commonly applied to predict groundwater yield potential. Conditioning factors should be considered depending on regional characteristics. For this reason, the correlation between the factors and groundwater potential were analyzed preferentially through the frequency ratio model and the factors were selected; groundwater potential was estimated using 16 factors in this study. The 16 conditioning factors were constructed into a groundwater inventory, including nine topographic factors (convergence index, convexity, mass balance index (MBI), slope angle, slope height, topographic texture, topographic position index (TPI), topographic ruggedness index (TRI), and valley depth), two hydrological factors (flow path length, and slope length and steepness (LS)), forest type, soil material, land use, and two geological factors (lithology and distance from fault) (Figures 3 and 4). The conditioning factors were calculated and prepared using ArcGIS 10.3 software (ESRI, Redlands, CA, USA). Each dataset was converted into a grid format with 30-m spatial resolution for use in the groundwater inventory of the study area.

Topographic factors were calculated from a 1:5000 scale topographic map provided by the Korean National Geographic Information Institute. Spatial data, such as location and topography, were structured using ground control point measurements taken from digital aerial photographs and ground surveys. Aerial photographs were analyzed through numerical mapping, and further calibration was carried out through field surveys to create the topographic map. A digital elevation model (DEM) was first generated from the topographic map and then used to derive topographic factors, including convergence index, convexity, MBI, slope angle, slope height, topographic texture, TPI, TRI, and valley depth. Slope factor impacts groundwater recharge, with gentle slope areas having relatively high percolation and low surface runoff rates and steep areas having high surface runoff [48]. Soil moisture content is also related to slope, which affects precipitation direction [49]. Slope angle is strongly related to groundwater potential; therefore, groundwater-related topographic factors derived from DEM data with SAGA-GIS software [50] were used for modeling. Acceleration and deceleration, as well as flow convergence and divergence of flow, are mainly affected by the curvature of the area [51]. The hydrological factors flow path and LS factor were considered conditioning factors for hydrological features.

A forest map was also used, which was generated from field investigations and interpretation of aerial photographs. To construct the forest map, the near-infrared band was used for image analysis, in addition to the red-green-blue image. Moreover, soil material characteristics can impact the rate of surface water penetration into aquifers, which drives groundwater potential [52]. The soil material factor was extracted from a soil map published by the National Institute of Agricultural Sciences at 1:25,000 scale. Similarly, land cover has an impact on soil conditions such that storage and movement of groundwater change when land cover changes; the land use factor was extracted from a digital land cover map provided by the Korea Ministry of Environment at 1:25,000 scale. Land use maps were classified into 22 medium-level categories through application of automatic image classification to aerial photographs, and the accuracy was enhanced using additional high-resolution satellite images from KOMPSAT-2 and 3. The land cover map was reclassified into seven land cover categories: urban, farmland, forest, grassland, wetland, bare land, and water.

Geological factors, including lithology and distance from a fault, were also considered in relation to groundwater characteristics. The lithology factor was extracted from a digital geological map produced by the Korea Institute of Geoscience and Mineral Resources at 1:50,000 scale. The study area was composed of 22 lithological units differing in lithology type and geological age. Distance from a fault was also calculated based on the geological map.

Table 1. Data layers describing groundwater potential.

Category	Factor	Scale	Data Type	Source Data (Year)
Pumping Test data	Specific capacity (SPC) Transmissivity (T)	-	Point	Field Survey (2008)
Topography	Convergence index Convexity Mass balance index (MBI) Slope angle Slope height Topographic texture Topographic position index (TPI) Topographic ruggedness index (TRI) Valley depth			Aerial Photography (2006–2016)
Hydrology	Flow path Slope length and steepness (LS) factor	1:50,000	Polygon	Field Survey (2008)
Forest	Forest type	1:25,000	Polygon	Aerial Photography, Field Survey (2004–2006)
Soil	Soil	1:25,000	Polygon	Aerial Photography, Field Survey (1998–2006)
Landcover	Landcover	1:5000	Polygon	Kompsat 2, 3 and Aerial Photography (2012)
Geology	Geology Distance from fault	1:250,000	Polygon	Aerial Photography, Field Survey (2004)

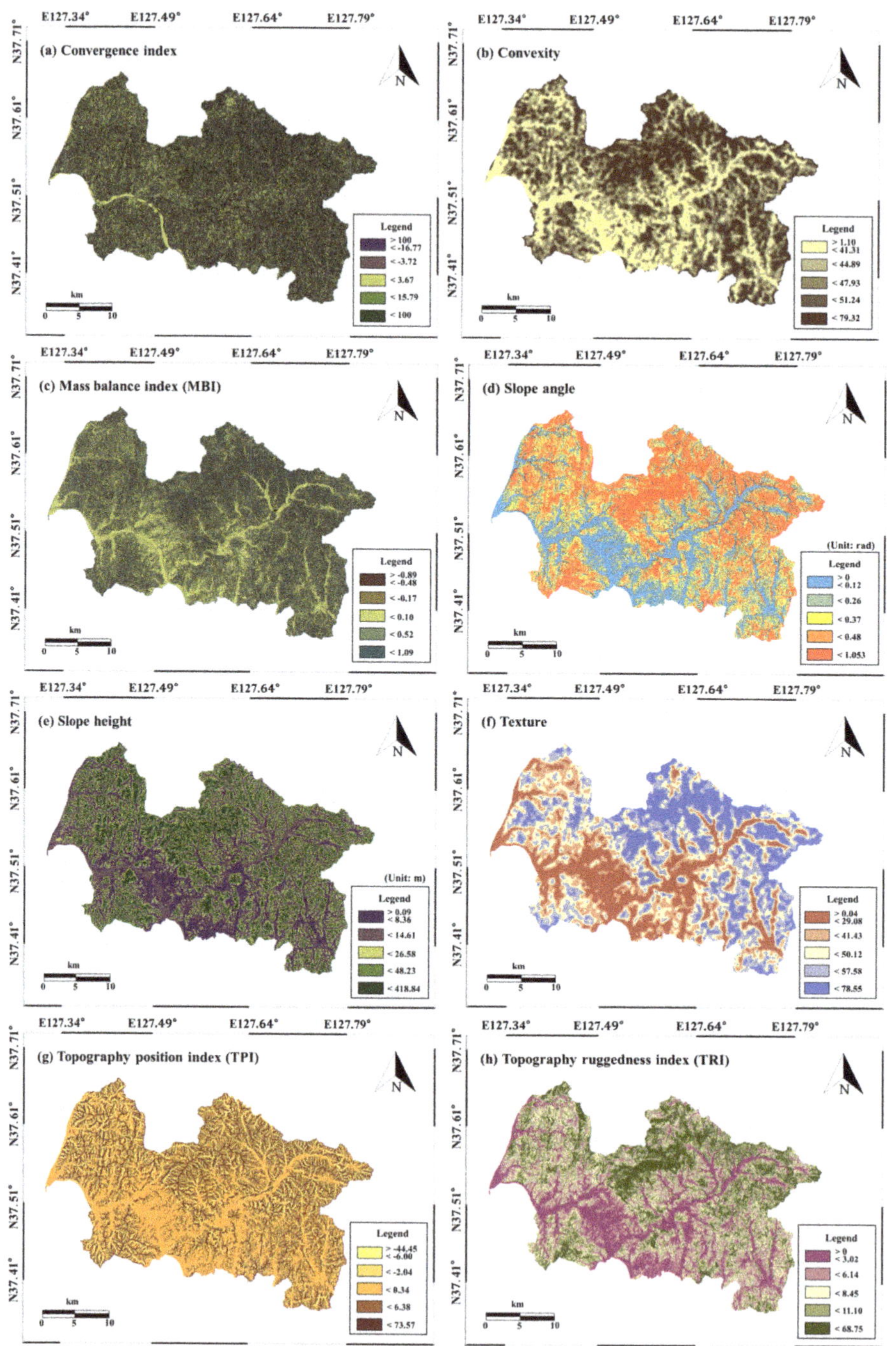

Figure 3. Groundwater conditioning factors I: (**a**) Convergence Index, (**b**) Convexity, (**c**) Mass balance index (MBI), (**d**) Slope angle, (**e**) Slope height, (**f**) Texture, (**g**) Topography position index (TPI) and (**h**) Topography ruggedness index (TRI).

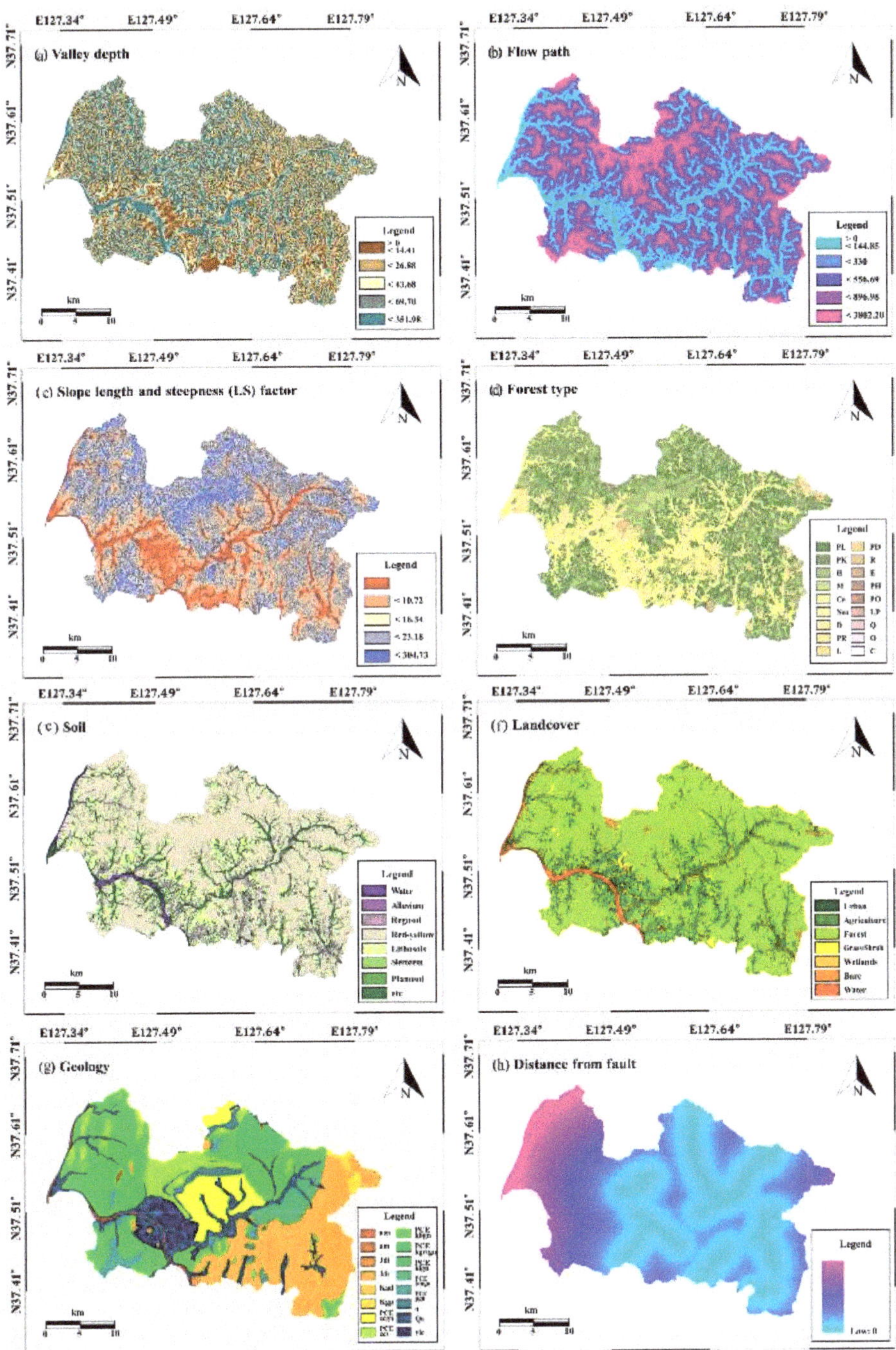

Figure 4. Groundwater conditioning factors II: (**a**) Valley depth, (**b**) Flow path, (**c**) Slope length and steepness (LS) factor, (**d**) Forest type, (**e**) Soil, (**f**) Land cover, (**g**) Geology and (**h**) Distance from fault.

4. Methodology

To be more specific, the purpose of this study was to map and test groundwater yield potential in Yangpyeong-gun, South Korea, using spatial data analysis in a GIS environment. This was performed by four main steps: First, groundwater yield data of specific capacity (SPC) and transmissivity (T) collected from 53 well locations were used. For the training data, 70% of each groundwater yield dataset was selected randomly, and FR and boosted tree (BT) models with classification were applied to the groundwater inventory using Statistica software (Dell Software, Aliso Viejo, CA, USA). Second, the inventory was constructed from nine topographic factors, two hydrological factors, forest type, soil material, land use, and two geological factors. All factors used in this study were generated and processed from remote sensing-based data, such as aerial photographs or imagery from KOMPSAT-2 and -3. Third, this study involved probabilistic analysis of FR, and two machine learning models: the boosted classification tree (BCT) and FR-BCT ensemble models, which were applied to groundwater yield data. Comparative analysis was conducted to compare the models used in this study. Finally, to quantitatively evaluate the performance of the models, the receiver operating characteristics (ROC) and area under the curve (AUC) were used. The study was conducted, as shown in Figure 5.

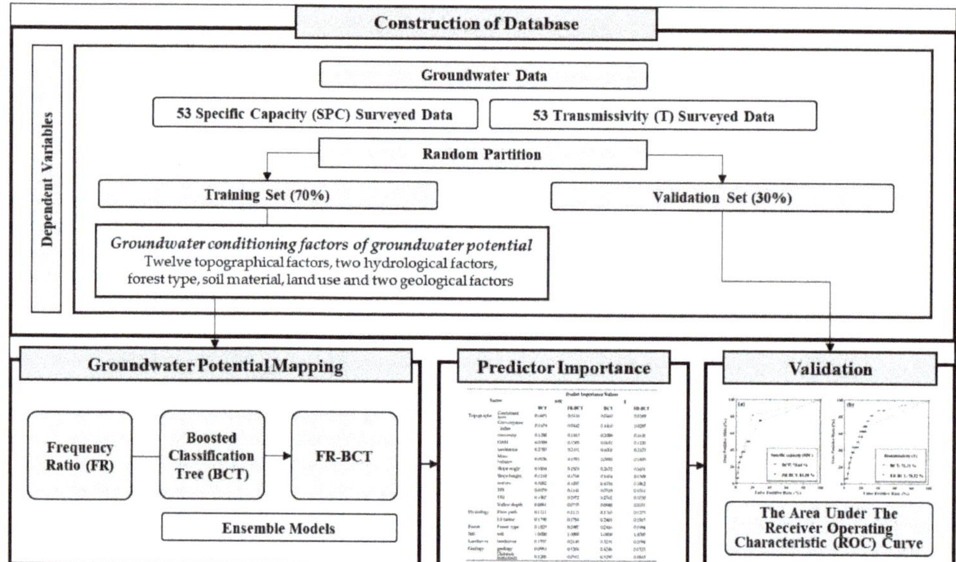

Figure 5. Data flow of this study.

4.1. Frequency Ratio (FR) Model

FR is an effective stochastic method for evaluating the effects of various factors on the occurrence of a particular event [53]. Thus, the FR value represents the ratio of occurrence of a particular event to the area ratio for each class [54]. A larger FR value represents a stronger relationship between the probability of occurrence and the specific variable [55,56]. This method allows for the clear and simple analysis of the relationship of each factor to the event [57].

To carry out spatial FR analysis, factors related to groundwater potential were classified into ten classes. Among numerous available classification techniques, factors in this study were classified using the quantile technique, which divides classes into equal areas. FR values were calculated using training data for each factor. Each class of each modulator was weighted. Higher FR values represent a stronger relationship between the class of each factor and groundwater potential, whereas for lower FR values, the effect of the class of each factor on groundwater potential is small. If FR is greater than 1, the effect is significant; if FR is less than 1, the effect is not significant [56]. To construct a groundwater potential map using FR to represent the relative magnitude of the groundwater potential, the FR values calculated for each factor were determined as follows:

$$FR = \frac{P_{trn}}{P_{total}} \tag{4}$$

where P_{trn} is the ratio of the number of SPC data points above a certain level and P_{total} indicates the ratio of the number of pixels in a certain class to the total number of pixels in the study area. A greater FR value for potential indicates higher groundwater potential; a lower value indicates a lower groundwater potential. In this study, FR values for each conditioning factor were used to weight the ensemble FR-BCT model.

4.2. Boosted Classification Tree

In recent years, decision tree models have been used in various fields as a machine learning method [58], including for groundwater potential mapping [52]. Decision tree models perform attribute tests on non-terminal nodes to represent the results on the terminal node, using a tree-like hierarchy that constructs a classification tree of a simple structure [59]. One of the benefits of this method is that the classification process can be graphically represented. However, the results cannot be formed into multiple outputs and the performance of the model depends on the type of data. Many algorithms have been developed from decision trees: classification and regression tree [60], chi-square automatic interaction detector decision tree [61], Iterative Dichotomiser 3 [62], and J48 (C4.5 decision tree) [63]. In addition, ensemble models using sequences of classifiers have been widely developed. Representative ensemble methods such as voting, bagging (sub-sampling), and boosting have also been applied to the decision tree method, including BT algorithms. Therefore, in this study, representative decision tree algorithms of BT models were used to compare the performance of each model's groundwater potential modeling and prediction accuracy.

The BT model is a tree-based machine learning model using the stochastic gradient boosting method. In the last few years, this algorithm has become one of the most powerful machine learning techniques used for prediction. In the BT algorithm, continuous or categorical input factors can be used for classification and regression problems [64].

The BT algorithm is implemented by applying a boosting method to the regression tree. The basic method involves calculating a simple tree sequence in which each successive tree is built against the prediction residual of the preceding tree. This method creates two trees of data for two samples at each split node. Even if the relationship between predictive and dependent variables is nonlinear, the weighting of such trees can support high accuracy of the predicted value. Thus, the gradient boosting method for weighted expansion of simple trees is one of the most common and powerful machine learning algorithms.

All machine learning algorithms are prone to overfitting, which involves a good fit for learning data but a lack of improvement in the predictability of each model. In other words, this is a common problem that applies to most algorithms used for predictive machine learning. A common solution to this problem is to evaluate the quality of the model fit by predicting observations from test samples of "used" data before evaluating each model [65,66]. The accuracy of each solution can be measured in this way to determine when the overflow occurred.

To overcome this difficulty, which is a major problem facing most machine learning algorithms used in predictive models, a specific approach was selected for the BT models. A continuous simple tree is generated using only subsamples selected randomly from the entire dataset. That is, each successive tree is created for the predicted residuals of an independently extracted random sample. Randomness can be added to any degree to protect against overfitting and can provide good predictability. Continuous boosting calculations for independently sampled input samples are known as probabilistic gradient boosting techniques.

4.3. Ensemble Modelling

Using the two methodologies described above, ensemble methods of FR and BCT were applied in this study. The probabilistic method FR was used to assess the impact of all types of regulatory factors and assign appropriate weights to each class according to their impact on groundwater yield. Using the FR method, individual weights were derived for each factor. Each conditioning coefficient was then reclassified using the derived weight values, and the reclassified dataset was analyzed using the BCT tree-based machine learning models. Finally, a groundwater potential map was constructed using the BCT and FR-BCT ensemble techniques for comparative analysis.

4.4. Assessment on Model Performance

The performance of groundwater potential classification was assessed using two statistical indicators: sensitivity and specificity. Sensitivity is the percentage of correctly classified pixels in areas with high groundwater potential; specificity is the percentage of pixels classified as having a low groundwater potential. Sensitivity and specificity are calculated as follows [67]:

$$Sensitivity = \frac{TP}{TP + FN}, \tag{5}$$

$$Specificity = \frac{TN}{FP + TN}, \tag{6}$$

The numbers of correctly classified pixels are denoted as true positives (TP) and true negatives (TN). Conversely, the numbers of misclassified pixels are expressed as false positives (FP) and false negatives (FN).

In this study, ROC curves were used to evaluate the overall performance of the groundwater potential model. The ROC curve has been applied in various fields as a standard method for evaluating the general performance of a model [68]. This curve is plotted using sensitivity as the x-axis and 100 − specificity as the y-axis. The general performance of the model can be quantitatively assessed based on the AUC value, representing the area under the ROC curve. AUC values range from 0.5 to 1. A value of 0.5 represents a model with very low accuracy. In contrast, 1 represents a perfect model with the highest possible accuracy, and an AUC close to 1 indicates good performance. Generally, when the AUC value is greater than 0.8, the model shows adequate performance [69].

5. Results

5.1. Results from the Frequency Ratio Model

Table A1 presents the correlations of FR values between groundwater data (SPC or T) and groundwater conditioning factors derived from the FR model. The FR is a representative value of the statistical proportional position of well locations with SPC values above a specific level. Correlation between groundwater well data and each factor could be shown from the distribution of values biased according to each class. Areas with high FR values are of great importance for groundwater management because they have high groundwater potential. The characteristics of land cover in the area of this study are high in forest area and agricultural area, and relatively low in urban area. Although there are many groundwater wells in urban areas, the urban area is mixed with rural areas, so it requires a different approach from metropolis.

The topographic factor convexity showed a strong correlation with groundwater potential in the 1.1–43.19 class for FR values of over 1.89 and 2.63 for SPC and T, respectively. Similarly, MBI showed a high correlation with SPC (2.16) and T (1.84) in the -0.33 to 0.1 class. The highest FR values of 4.32 for SPC and 4.21 for T were observed when the slope angle was greater than 0 m and less than 0.05 m, indicating that this factor is strongly correlated with groundwater potential. FR values tended to decrease with increasing slope angle and slope height. For topographic texture, the 0.04–29.08 class exhibited the highest FR values with SPC (2.97) and T (3.95). Low flow path values also led to FR values over 1, indicating that this factor was correlated with groundwater potential.

Among land cover types, urban area showed the strongest relationship with groundwater potential (SPC: 6.66; T: 7.92), followed by wetlands. These results could also be interpreted as showing that the use frequency of wells in urban areas is high. Meanwhile, distance from a fault had FR values of 2.16 for SPC and 3.16 for T in the 0–530.75 class. Among geological factors, alluvium showed a strong correlation with the groundwater data (SPC: 2.93; T: 3.80), followed by granite porphyry (SPC: 1.45; T: 1.01).

5.2. Construction of Groundwater Potential Maps

The groundwater potential map was modeled using training datasets of SPC and T. The performance of a groundwater potential model depends on the selection of factors. The groundwater potential map was constructed by training the groundwater potential model. First, a groundwater potential value was generated for each pixel in Yangpyeong-gun. Each pixel was indexed by its predicted groundwater potential value. The results of groundwater potential were reclassified using the 1.0 standard deviation method, which is based on the distribution of individual values in the results for each model. In the groundwater potential map, areas with high (low) groundwater potential are shaded red (blue) (Figure 6). All models showed similar distributions of groundwater potential, and the north, southwest, and southeast areas surrounding the central valley region of the study area all showed low potential.

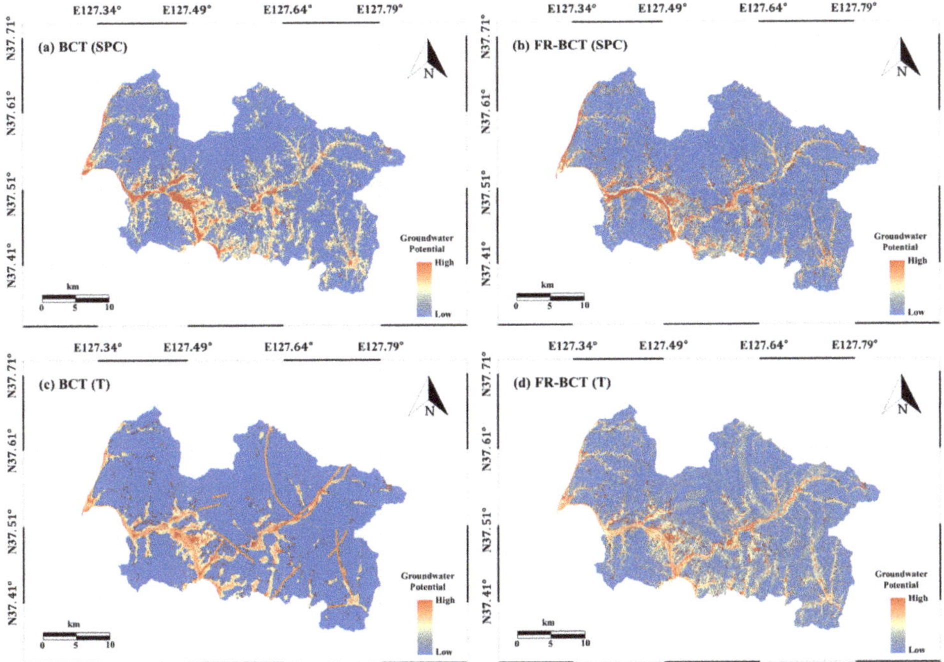

Figure 6. Groundwater potential maps based on (**a**) boosted classification tree (BCT) and (**b**) frequency ratio (FR)-BCT models with specific capacity (SPC) data, and (**c**) BCT and (**d**) FR-BCT models with transmissivity (T) data.

Furthermore, the predictor importance values of each factor were calculated from the BCT modeling results by summing the decreases in node-impurity values (Table 2). All predictor importance values were scaled to a maximum of 1.0, as the value assigned to the largest sum among all factors, indicating the most strongly related factor, relatively. For both SPC and T, soil showed the highest predictor importance values in all models, with a value of 1.0. Topographic texture was the second most important factor in the BCT models, with values of 0.3101 and 0.4206, for SPC and T data, respectively. Meanwhile, FR-BCT models showed that forest type and land cover were the second strongest predictors, with importance values of 0.1704 and 0.2295 for SPC and T data, respectively. The importance of TPI, MBI, and valley depth were low in all FR models; convergence index, valley depth, and distance from a fault fell into the third lowest positions based on the FR-BCT models.

Table 2. Predictor importance values of each factor for the BCT and FR-BCT models.

| | | Predictor Importance Values | | | |
| | | SPC | | T | |
Factor		BCT	FR-BCT	BCT	FR-BCT
Topography	Convergence index	0.1689	0.0400	0.1518	0.0494
	convexity	0.1285	0.1223	0.1913	0.1698
	Mass balance index (MBI)	0.0566	0.1345	0.0703	0.1680
	Slope angle	0.1909	0.1443	0.2734	0.1850
	Slope height	0.1245	0.1393	0.1711	0.1750
	Topographic texture	0.3101	0.1196	0.4206	0.1975
	Topographic position index (TPI)	0.0387	0.0990	0.0565	0.1246
	Topographic ruggedness index (TRI)	0.1967	0.1658	0.2917	0.2003
	Valley depth	0.0887	0.0696	0.0929	0.0513
Hydrology	Flow path	0.1123	0.0917	0.1819	0.1407
	Slope length and steepness (LS) factor	0.1835	0.1466	0.2747	0.1935
Forest	Forest type	0.1821	0.1704	0.2084	0.1694
Soil	Soil	1.0000	1.0000	1.0000	1.0000
Landcover	Landcover	0.1946	0.1572	0.3285	0.2295
Geology	geology	0.1002	0.0996	0.1619	0.1386
	Distance from fault	0.1271	0.0497	0.3105	0.1061

5.3. Model Performance Evaluation

In this study, the groundwater potential model was evaluated based on statistical indices; AUC was used to quantitatively assess the mapping accuracy. As aforementioned, testing was performed based on the 30% of the groundwater well data collected by field investigation; and since groundwater has less seasonal change than surface water, this study did not consider seasonal change for groundwater. Figure 7 presents the model accuracy rate for the SPC (BCT model: 80.48%; FR-BCT model: 87.75%) and T (BCT model: 72.27%; FR-BCT model: 81.49%) well data. In general, all groundwater potential mapping results and modeling of groundwater potential showed good performance; however, the ensemble models showed improved accuracy by approximately 6%. Figure 7 also shows the performance of the groundwater potential models using the ROC curve method. All groundwater potential models performed well in terms of groundwater potential evaluation results (AUC > 0.7). The testing results of the BCT ensemble model show that 20% of the groundwater potential area includes approximately 80% of the valid groundwater wells for SPC, whereas the testing results of the ensemble model for T show that 30% of the groundwater area includes over 80% of the valid groundwater wells. Compared to groundwater potential mapping with the single machine learning model, BCT, all groundwater potential models using the ensemble method with both FR and BCT showed better performance, with 7.27% and 9.22% higher accuracy, respectively, than the BCT model alone. The difference in AUC results showed that the ensemble model provided better results than the individual modeling process.

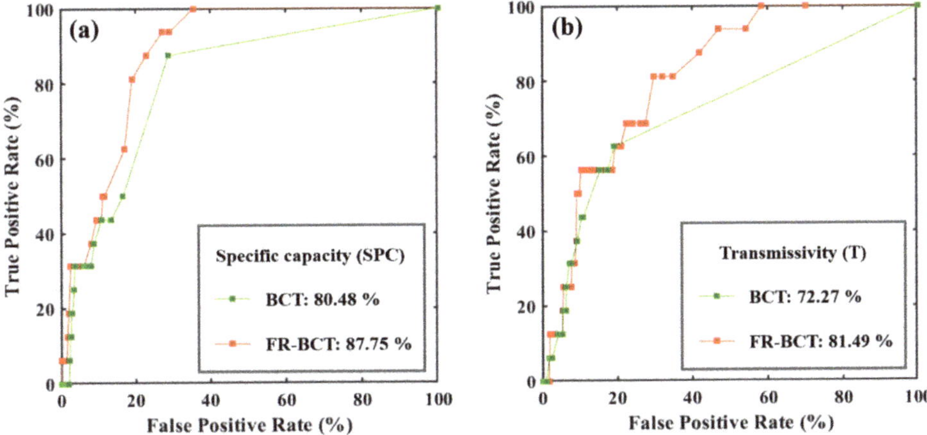

Figure 7. Testing results of the BCT and FR-BCT models for (**a**) SPC and (**b**) T groundwater data.

6. Discussion

In this paper, the relationship between conditioning factors and groundwater was first analyzed through the stochastic method of FR. By applying the ensemble technique to the BCT model based on the stochastic weighting, it showed effectiveness in the study of groundwater with high uncertainty. In terms of data, this study was based on data created by governments and public institutions and released to the public; at the same time, it is bound by limitations in data collection. Since the importance of data used for training in data-based learning is very high, model accuracy will be improved if more well data is used in future studies.

Few case studies have applied ensemble models from machine learning algorithms in South Korea. The results of this study confirm that the performance of a groundwater potential model can be improved using an existing probability model and machine learning ensemble. Model performance was evaluated based on the ROC, and the prediction rate of the BCT model showed an improvement of 6.1% with FR-BCT for SPC and 6.0% for T compared to the single machine learning model, BCT, indicating that the ensemble method greatly improved model performance. This improvement occurred because the ensemble model could reduce bias using the BT model and improve its predictive ability by avoiding the overfitting problem of basic classification [70]. This finding is consistent with other studies that concluded that the predictive performance of models was improved with a machine learning ensemble model [71].

Remote sensing is a powerful data source that is widely used for monitoring environmental issues; however, since groundwater does not exist on the surface, groundwater can only be indirectly estimated by using remote sensing. Heretofore, many studies have attempted to reduce the uncertainty of groundwater spatially. As a result of applying the proposed FR-BCT model with existing probability models and the machine learning method of the BCT model, the accuracy was relatively improved or similar to previous studies [3,25,34,68]. In addition, by showing accuracy improvements in single and composite models, it has shown potential for reducing the uncertainty of groundwater potential mapping.

7. Conclusions

The modern global water shortage requires effective water management and planning. Indiscreet use of water resources and inadequate water management can disrupt the continuous and reliable supply of water. The first step in properly planning water resource usage is to accurately predict and respond to the current status of critical resources. Groundwater represents an excellent water source, especially in water-scarce regions. However, the uncertainty of groundwater availability is high; therefore, estimation of groundwater potential is essential. Mapping of groundwater potential is an essential challenge facing effective groundwater resource management and conservation planning.

Various methods of groundwater potential mapping have been proposed. Improvement of the groundwater potential model is one method for estimating the uncertainty of a groundwater model. Although new machine learning technologies are continually improving in predictive performance, not all methods can be effectively applied in areas where data are scarce, because it may not be possible to generalize from a small labeled dataset. Therefore, FR analysis and the BCT model were applied along with the proposed FR-BCT model, which is an ensemble model of these two machine learning models. For this purpose, 16 groundwater control factors based on remote-sensing data were applied to the models: nine topographic factors, two hydrological factors, forest type, soil material, land use, and two geological factors. The model was trained and tested using groundwater well data; 53 wells were separated into training (70%) and testing (30%) datasets. The proposed FR-BCT model was compared with existing probability models and the machine learning method of the BCT model.

These results are useful for supporting comprehensive management of groundwater exploration and groundwater recharge. The method used in this study can be applied to other areas reliant on groundwater use. Managers and policymakers can effectively analyze groundwater potential modeling results to maximize the benefits of management. However, further testing is required in other research areas to determine how reliably the proposed ensemble model reflects groundwater potential.

Author Contributions: Conceptualization, S.L. and M.-J.L.; Data curation, S.L. and S.L.; Formal analysis, Y.H.; Investigation, Y.H. and S.L.; Methodology, S.L. and Y.H.; Project administration, M.-J.L.; Resources, S.L.; Software, S.L. and S.L.; Supervision, M.-J.L.; Validation, S.L.; Visualization, S.L.; Writing—riginal draft, S.L.; Writing—review and editing, M.-J.L. All authors have read and agreed to the published version of the manuscript.

Funding: This research was conducted at the Korea Environment Institute (KEI) with support from the Basic Science Research Program through the National Research Foundation of Korea (NRF) funded by the Ministry of Education (NRF-2018R1D1A1B07041203). This research was conducted by the Basic Research Project of the Korea Institute of Geoscience and Mineral Resources (KIGAM) funded by the Ministry of Science and ICT. This research was also conducted with support from 'Future Forecasting Government Management Support: A Study on the Modeling of Climate and Environment for the Transition to Data Science' funded by the National Research Council for Economics, Humanities and Social Sciences.

Conflicts of Interest: The authors declare no conflict of interest.

Appendix A

Table A1. Results of the frequency ratio model.

Factor	Class	No. of SPC	% of SPC	No. of T	% of T	No. of Pixels in Domain	% of Pixels in Domain	Frequency Ratio of SPC	Frequency Ratio of T
Convergence index	−100–−29.9	7	18.92	6	16.22	97,261	10.00	1.89	1.62
	−29.9–−16.77	2	5.41	2	5.41	97,262	10.00	0.54	0.54
	−16.77–−8.98	5	13.51	4	10.81	97,261	10.00	1.35	1.08
	−8.98–−3.72	5	13.51	5	13.51	97,262	10.00	1.35	1.35
	−3.72–0	5	13.51	3	8.11	89,697	9.22	1.47	0.88
	0–3.67	4	10.81	6	16.22	104,826	10.78	1.00	1.50
	3.67–8.59	2	5.41	2	5.41	97,261	10.00	0.54	0.54
	8.59–15.79	2	5.41	4	10.81	97,262	10.00	0.54	1.08
	15.79–28.28	2	5.41	3	8.11	97,261	10.00	0.54	0.81
	28.28–100	3	8.11	2	5.41	97,262	10.00	0.81	0.54
Convexity	1.1–38.72	12	32.43	9	24.32	97,260	10.00	3.24	2.43
	38.72–41.31	7	18.92	12	32.43	97,263	10.00	1.89	3.24
	41.31–43.19	11	29.73	10	27.03	97,261	10.00	2.97	2.70
	43.19–44.89	1	2.70	1	2.70	97,262	10.00	0.27	0.27
	44.89–46.43	3	8.11	4	10.81	97,258	10.00	0.81	1.08
	46.43–47.93	0	0.00	0	0.00	97,263	10.00	0.00	0.00
	47.93–49.47	0	0.00	0	0.00	97,263	10.00	0.00	0.00
	49.47–51.24	1	2.70	0	0.00	97,262	10.00	0.27	0.00
	51.24–53.52	1	2.70	1	2.70	97,260	10.00	0.27	0.27
	53.52–79.32	1	2.70	0	0.00	97,263	10.00	0.27	0.00

Table A1. *Cont.*

Factor	Class	No. of SPC	% of SPC	No. of T	% of T	No. of Pixels in Domain	% of Pixels in Domain	Frequency Ratio of SPC	Frequency Ratio of T
Mass balance index (MBI)	−0.89−−0.62	1	2.70	0	0.00	97,261	10.00	0.27	0.00
	−0.62−−0.48	1	2.70	0	0.00	97,262	10.00	0.27	0.00
	−0.48−−0.33	0	0.00	1	2.70	97,261	10.00	0.00	0.27
	−0.33−−0.17	9	24.32	7	18.92	97,262	10.00	2.43	1.89
	−0.17−−0.02	10	27.03	10	27.03	97,261	10.00	2.70	2.70
	−0.02−0.1	8	21.62	12	32.43	97,262	10.00	2.16	3.24
	0.1−0.33	3	8.11	3	8.11	97,261	10.00	0.81	0.81
	0.33−0.52	3	8.11	2	5.41	97,262	10.00	0.81	0.54
	0.52−0.68	1	2.70	0	0.00	97,261	10.00	0.27	0.00
	0.68−1.09	1	2.70	2	5.41	97,262	10.00	0.27	0.54
Slope angle (rad)	0−0.05	16	43.24	16	43.24	97,261	10.00	4.32	4.32
	0.05−0.12	8	21.62	11	29.73	97,262	10.00	2.16	2.97
	0.12−0.2	5	13.51	4	10.81	97,261	10.00	1.35	1.08
	0.2−0.26	3	8.11	3	8.11	97,262	10.00	0.81	0.81
	0.26−0.32	3	8.11	1	2.70	97,261	10.00	0.81	0.27
	0.32−0.37	0	0.00	0	0.00	97,262	10.00	0.00	0.00
	0.37−0.42	0	0.00	1	2.70	97,261	10.00	0.00	0.27
	0.42−0.48	1	2.70	1	2.70	97,262	10.00	0.27	0.27
	0.48−0.56	1	2.70	0	0.00	97,261	10.00	0.27	0.00
	0.56−1.05	0	0.00	0	0.00	97,262	10.00	0.00	0.00
Slope height (m)	0.09−6.39	4	10.81	2	5.41	97,261	10.00	1.08	0.54
	6.39−8.36	13	35.14	13	35.14	97,262	10.00	3.51	3.51
	8.36−11.05	11	29.73	11	29.73	97,261	10.00	2.97	2.97
	11.05−14.61	4	10.81	6	16.22	97,262	10.00	1.08	1.62
	14.61−19.86	1	2.70	2	5.41	97,261	10.00	0.27	0.54
	19.86−26.58	3	8.11	2	5.41	97,262	10.00	0.81	0.54
	26.59−35.51	0	0.00	0	0.00	97,261	10.00	0.00	0.00
	35.51−48.23	0	0.00	1	2.70	97,262	10.00	0.00	0.27
	48.23−71.03	0	0.00	0	0.00	97,261	10.00	0.00	0.00
	71.03−418.84	1	2.70	0	0.00	97,262	10.00	0.27	0.00
Topographic texture	0.04−18.87	11	29.73	14	37.84	97,261	10.00	2.97	3.78
	18.87−29.08	11	29.73	15	40.54	97,259	10.00	2.97	4.05
	29.08−36.08	7	18.92	2	5.41	97,264	10.00	1.89	0.54
	36.08−41.43	4	10.81	3	8.11	97,261	10.00	1.08	0.81
	41.43−46.05	1	2.70	1	2.70	97,262	10.00	0.27	0.27
	46.05−50.12	0	0.00	0	0.00	97,262	10.00	0.00	0.00
	50.12−53.76	1	2.70	0	0.00	97,261	10.00	0.27	0.00
	53.76−57.58	1	2.70	1	2.70	97,262	10.00	0.27	0.27
	57.58−62.07	1	2.70	1	2.70	97,261	10.00	0.27	0.27
	62.07−78.55	0	0.00	0	0.00	97,262	10.00	0.00	0.00
Topographic position index (TPI)	−44.45−−9.66	1	2.70	0	0.00	97,261	10.00	0.27	0.00
	−9.66−−6	2	5.41	2	5.41	97,262	10.00	0.54	0.54
	−6−−3.74	6	16.22	4	10.81	97,261	10.00	1.62	1.08
	−3.74−−2.04	2	5.41	2	5.41	97,262	10.00	0.54	0.54
	−2.04−−0.68	6	16.22	8	21.62	97,261	10.00	1.62	2.16
	−0.68−0.34	11	29.73	14	37.84	97,262	10.00	2.97	3.78
	0.34−2.78	4	10.81	4	10.81	97,261	10.00	1.08	1.08
	2.78−6.38	4	10.81	1	2.70	97,262	10.00	1.08	0.27
	6.38−11.46	0	0.00	1	2.70	97,261	10.00	0.00	0.27
	11.46−73.57	1	2.70	1	2.70	97,262	10.00	0.27	0.27
Topographic ruggedness index (TRI)	0−1.16	17	45.95	18	48.65	97,261	10.00	4.59	4.86
	1.16−3.02	7	18.92	10	27.03	97,262	10.00	1.89	2.70
	3.02−4.78	3	8.11	3	8.11	97,261	10.00	0.81	0.81
	4.78−6.14	6	16.22	3	8.11	97,262	10.00	1.62	0.81
	6.14−7.31	2	5.41	1	2.70	97,261	10.00	0.54	0.27
	7.31−8.45	0	0.00	0	0.00	97,262	10.00	0.00	0.00
	8.45−9.66	1	2.70	1	2.70	97,261	10.00	0.27	0.27
	9.66−11.1	1	2.70	0	0.00	97,261	10.00	0.27	0.00
	11.1−13.06	0	0.00	1	2.70	97,262	10.00	0.00	0.27
	13.06−68.75	0	0.00	0	0.00	97,262	10.00	0.00	0.00
Valley depth	0−9.29	2	5.41	1	2.70	97,261	10.00	0.54	0.27
	9.29−14.41	5	13.51	3	8.11	97,262	10.00	1.35	0.81
	14.41−20.19	0	0.00	2	5.41	97,261	10.00	0.00	0.54
	20.19−26.88	2	5.41	3	8.11	97,262	10.00	0.54	0.81
	26.88−34.55	5	13.51	7	18.92	97,261	10.00	1.35	1.89
	34.55−43.68	6	16.22	5	13.51	97,262	10.00	1.62	1.35
	43.68−54.86	3	8.11	6	16.22	97,261	10.00	0.81	1.62
	54.86−69.7	3	8.11	3	8.11	97,262	10.00	0.81	0.81
	69.7−93.65	4	10.81	1	2.70	97,261	10.00	1.08	0.27
	93.65−351.98	6	15.79	6	16.22	97,262	10.00	1.89	1.62

Table A1. *Cont.*

Factor	Class	No. of SPC	% of SPC	No. of T	% of T	No. of Pixels in Domain	% of Pixels in Domain	Frequency Ratio of SPC	Frequency Ratio of T
Flow path	0–60	12	32.43	12	32.43	93,864	9.65	3.36	3.36
	72.42–144.85	6	16.22	5	13.51	100,271	10.31	1.57	1.31
	150–229.7	7	18.92	8	21.62	92,423	9.50	1.99	2.28
	234.85–330	4	10.81	4	10.81	101,044	10.39	1.04	1.04
	332.13–434.55	3	8.11	4	10.81	95,153	9.78	0.83	1.11
	434.55–556.69	4	10.81	3	8.11	100,626	10.35	1.04	0.78
	556.69–704.55	1	2.70	1	2.70	96,887	9.96	0.27	0.27
	704.55–896.98	0	0.00	0	0.00	97,622	10.04	0.00	0.00
	896.98–1193.96	0	0.00	0	0.00	96,752	9.95	0.00	0.00
	1193.97–3802.2	0	0.00	0	0.00	97,973	10.07	0.00	0.00
LS factor	0–1.6	14	37.84	16	43.24	97,261	10.00	3.78	4.32
	1.6–4.63	10	27.03	11	29.73	97,262	10.00	2.70	2.97
	4.63–7.78	5	13.51	5	13.51	97,261	10.00	1.35	1.35
	7.78–10.72	4	10.81	2	5.41	97,262	10.00	1.08	0.54
	10.72–13.52	1	2.70	1	2.70	97,261	10.00	0.27	0.27
	13.52–16.34	1	2.70	1	2.70	97,262	10.00	0.27	0.27
	16.34–19.42	1	2.70	0	0.00	97,261	10.00	0.27	0.00
	19.42–23.18	1	2.70	1	2.70	97,262	10.00	0.27	0.27
	23.18–29.11	0	0.00	0	0.00	97,261	10.00	0.00	0.00
	29.11–304.73	0	0.00	0	0.00	97,262	10.00	0.00	0.00
Forest type	Deciduous pine tree (PL)	3	8.11	3	8.11	240,329	24.71	0.33	0.33
	Pine forest (PK)	2	5.41	1	2.70	133,782	13.75	0.39	0.20
	Broadleaved forest (H)	0	0.00	0	0.00	181,244	18.63	0.00	0.00
	Mixed forest of soft and hardwood (M)	0	0.00	1	2.70	51,415	5.29	0.00	0.51
	Chestnut forest (Ca)	1	2.70	0	0.00	3168	0.33	8.30	0.00
	Non–forest (ND)	28	75.68	30	81.08	241,742	24.85	3.04	3.26
	Pine forest (D)	0	0.00	0	0.00	20,816	2.14	0.00	0.00
	Pinus rigida forest (PR)	3	8.11	2	5.41	75,341	7.75	1.05	0.70
	Farmland (L)	0	0.00	0	0.00	5653	0.58	0.00	0.00
	Needleleaf artificial forest (PD)	0	0.00	0	0.00	9602	0.99	0.00	0.00
	Left–over area (R)	0	0.00	0	0.00	6189	0.64	0.00	0.00
	Dentuded land (E)	0	0.00	0	0.00	119	0.01	0.00	0.00
	Broadleaved artificial forest (PH)	0	0.00	0	0.00	1664	0.17	0.00	0.00
	Poplar forest (Po)	0	0.00	0	0.00	244	0.03	0.00	0.00
	Grassland (LP)	0	0.00	0	0.00	877	0.09	0.00	0.00
	Oak forest (Q)	0	0.00	0	0.00	225	0.02	0.00	0.00
	Fine–grained wood (O)	0	0.00	0	0.00	31	0.00	0.00	0.00
	Coniferous forest (C)	0	0.00	0	0.00	174	0.02	0.00	0.00
Soil	Water	4	10.81	2	5.41	13,501	1.39	7.79	3.89
	Alluvium	7	18.92	8	21.62	97,462	10.02	1.89	2.16
	Regosol	2	5.41	2	5.41	76,862	7.90	0.68	0.68
	Red–yellow	4	10.81	6	16.22	107,414	11.04	0.98	1.47
	Lithosols	11	29.73	10	27.03	600,412	61.73	0.48	0.44
	Sierozem	7	18.92	7	18.92	60,429	6.21	3.05	3.05
	Planosol	1	2.70	1	2.70	245	0.03	107.29	107.29
	Other	1	2.70	1	2.70	16,290	1.67	1.61	1.61
Landcover	Urban	9	24.32	11	29.73	35,546	3.65	6.66	8.13
	Agriculture	15	40.54	14	37.84	179,578	18.46	2.20	2.05
	Forest	9	24.32	7	18.92	703,777	72.36	0.34	0.26
	Grass/Shrub	1	2.70	1	2.70	14,721	1.51	1.79	1.79
	Wetlands	1	2.70	1	2.70	4522	0.46	5.81	5.81
	Bare	0	0.00	0	0.00	13,032	1.34	0.00	0.00
	Water	2	5.41	3	8.11	21,465	2.21	2.45	3.67
Geology	Non	1	2.70	1	2.70	11,000	1.13	2.39	2.39
	Alluvium (Qa)	12	32.43	15	40.54	107,636	11.07	2.93	3.66
	Ganite–bearing granitic gneiss (PCEkgrtgn)	0	0.00	0	0.00	30,231	3.11	0.00	0.00
	Leucocratic gneiss (PCEklgn)	0	0.00	0	0.00	4314	0.44	0.00	0.00
	Migmatitic gneiss (PCEkmgn)	0	0.00	0	0.00	34,178	3.51	0.00	0.00
	Granite porphyry (Kgp)	0	0.00	0	0.00	9432	0.97	0.00	0.00
	Banded gneiss (PCEkbgn)	9	24.32	10	27.03	261,108	26.85	0.91	1.01
	Schists (PCEccs)	0	0.00	0	0.00	74,084	7.62	0.00	0.00
	Porphyroblastic gneiss (PCEpgn)	0	0.00	0	0.00	38,778	3.99	0.00	0.00
	Amphibolite (am)	0	0.00	0	0.00	1261	0.13	0.00	0.00
	Granite porphyry (Jgr)	14	37.84	10	27.03	253,162	26.03	1.45	1.04
	Quartzite (Q)	0	0.00	0	0.00	25,512	2.62	0.00	0.00

Table A1. *Cont.*

Factor	Class	No. of SPC	% of SPC	No. of T	% of T	No. of Pixels in Domain	% of Pixels in Domain	Frequency Ratio of SPC	Frequency Ratio of T
	Yangpyeong Igneous Complex (yic)	1	2.70	1	2.70	40,910	4.21	0.64	0.64
	Gneiss (PCEccgn)	0	0.00	0	0.00	73,955	7.60	0.00	0.00
	Diorite (Jdi)	0	0.00	0	0.00	3383	0.35	0.00	0.00
	Acidic (kad)	0	0.00	0	0.00	3671	0.38	0.00	0.00
Distance from fault	0–530.75	8	21.62	12	32.43	97,236	10.00	2.16	3.24
	531.6–1081.66	1	2.70	1	2.70	96,925	9.97	0.27	0.27
	1082.08–1611.36	1	2.70	3	8.11	97,419	10.02	0.27	0.81
	1612.2–2130.21	6	16.22	4	10.81	97,264	10.00	1.62	1.08
	2130.63–2673.2	3	8.11	4	10.81	97,368	10.01	0.81	1.08
	2674.21–3317.13	6	16.22	6	16.22	97,271	10.00	1.62	1.62
	3318.08–4440.4	3	8.11	0	0.00	97,341	10.01	0.81	0.00
	4440.91–7620.53	6	16.22	3	8.11	97,268	10.00	1.62	0.81
	7620.7–11187.71	2	5.41	2	5.41	97,259	10.00	0.54	0.54
	11187.91–17310.08	1	2.70	2	5.41	97,264	10.00	0.27	0.54

References

1. Oke, S.A.; Fourie, F. Guidelines to groundwater vulnerability mapping for Sub-Saharan Africa. *Groundw. Sustain. Dev.* **2017**, *5*, 168–177. [CrossRef]
2. Noh, C. Average Savings of 339 Reservoirs in Gyeonggi-do. Available online: http://www.todaykorea.co.kr/news/view.php?no=255887 (accessed on 8 September 2019).
3. Naghibi, S.A.; Pourghasemi, H.R.; Dixon, B. GIS-based groundwater potential mapping using boosted regression tree, classification and regression tree, and random forest machine learning models in Iran. *Environ. Monit. Assess.* **2016**, *188*, 44. [CrossRef] [PubMed]
4. Gaur, S.; Chahar, B.R.; Graillot, D. Combined use of groundwater modeling and potential zone analysis for management of groundwater. *Int. J. Appl. Earth Obs. Geoinf.* **2011**, *13*, 127–139. [CrossRef]
5. Abdulkareem, J.H.; Pradhan, B.; Sulaiman, W.N.A.; Jamil, N.R. Quantification of runoff as influenced by morphometric characteristics in a rural complex catchment. *Earth Syst. Environ.* **2018**, *2*, 145–162. [CrossRef]
6. Getirana, A.; Rodell, M.; Kumar, S.; Beaudoing, H.K.; Arsenault, K.; Zaitchik, B.; Save, H.; Bettadpur, S. GRACE improves seasonal groundwater forecast initialization over the US. *J. Hydrometeorol.* **2019**, *21*, 59–71. [CrossRef]
7. Li, B.; Rodell, M.; Kumar, S.; Beaudoing, H.K.; Getirana, A.; Zaitchik, B.F.; de Goncalves, L.G.; Cossetin, C.; Bhanja, S.; Mukherjee, A. Global GRACE data assimilation for groundwater and drought monitoring: Advances and challenges. *Water Resour. Res.* **2019**, *55*, 7564–7586. [CrossRef]
8. Nie, W.; Zaitchik, B.F.; Rodell, M.; Kumar, S.V.; Arsenault, K.R.; Li, B.; Getirana, A. Assimilating GRACE into a Land Surface Model in the presence of an irrigation-induced groundwater trend. *Water Resour. Res.* **2019**. [CrossRef]
9. Becker, M.W. Potential for satellite remote sensing of ground water. *Groundwater* **2006**, *44*, 306–318. [CrossRef] [PubMed]
10. Brunner, P.; Franssen, H.-J.H.; Kgotlhang, L.; Bauer-Gottwein, P.; Kinzelbach, W. How can remote sensing contribute in groundwater modeling? *Hydrogeol. J.* **2007**, *15*, 5–18. [CrossRef]
11. Balbarini, N.; Bjerg, P.L.; Binning, P.J.; Christiansen, A.V. Modelling Tools for Integrating Geological, Geophysical and Contamination Data for Characterization of Groundwater Plumes. Ph.D. Thesis, Department of Environmental Engineering, Technical University of Denmark, Kgs., Lyngby, Denmark, 2017.
12. Russoniello, C.; Michael, H.; Fernandez, C.; Andres, A.; He, C.; Madsen, J.A. *Investigation of Submarine Groundwater Discharge at Holts Landing State Park, Delaware: Hydrogeologic Framework, Groundwater Level and Salinity Observations*; Delaware Geological Survey, University of Delaware: Newark, DE, USA, 2017.
13. Helaly, A.S. Assessment of groundwater potentiality using geophysical techniques in Wadi Allaqi basin, Eastern Desert, Egypt–Case study. *Nriag J. Astron. Geophys.* **2017**, *6*, 408–421. [CrossRef]
14. Nampak, H.; Pradhan, B.; Manap, M.A. Application of GIS based data driven evidential belief function model to predict groundwater potential zonation. *J. Hydrol.* **2014**, *513*, 283–300. [CrossRef]

15. Ghorbani Nejad, S.; Falah, F.; Daneshfar, M.; Haghizadeh, A.; Rahmati, O. Delineation of groundwater potential zones using remote sensing and GIS-based data-driven models. *Geocarto Int.* **2017**, *32*, 167–187. [CrossRef]

16. Sameen, M.I.; Pradhan, B.; Lee, S. Self-learning random forests model for mapping groundwater yield in data-scarce areas. *Nat. Resour. Res.* **2019**, *28*, 757–775. [CrossRef]

17. Elfadaly, A.; Attia, W.; Lasaponara, R. Monitoring the environmental risks around Medinet Habu and Ramesseum Temple at West Luxor, Egypt, using remote sensing and GIS techniques. *J. Archaeol. Method Theory* **2018**, *25*, 587–610. [CrossRef]

18. Elmahdy, S.I.; Mohamed, M.M. Automatic detection of near surface geological and hydrological features and investigating their influence on groundwater accumulation and salinity in southwest Egypt using remote sensing and GIS. *Geocarto Int.* **2015**, *30*, 132–144. [CrossRef]

19. Fernandez, P.; Delgado, E.; Lopez-Alonso, M.; Poyatos, J.M. GIS environmental information analysis of the Darro River basin as the key for the management and hydrological forest restoration. *Sci. Total Environ.* **2018**, *613*, 1154–1164. [CrossRef]

20. Lee, J. Review of remote sensing studies on groundwater resources. *Korean J. Remote Sens.* **2017**, *33*, 855–866.

21. Hadžić, E.; Lazović, N.; Mulaomerović-Šeta, A. Application of mathematical models in defining optimal groundwater yield. *Procedia Environ. Sci.* **2015**, *25*, 112–119. [CrossRef]

22. Golkarian, A.; Naghibi, S.A.; Kalantar, B.; Pradhan, B. Groundwater potential mapping using C5. 0, random forest, and multivariate adaptive regression spline models in GIS. *Environ. Monit. Assess.* **2018**, *190*, 149. [CrossRef]

23. Kim, J.-C.; Jung, H.-S.; Lee, S. Groundwater productivity potential mapping using frequency ratio and evidential belief function and artificial neural network models: Focus on topographic factors. *J. Hydroinformatics* **2018**, *20*, 1436–1451. [CrossRef]

24. Rahmati, O.; Naghibi, S.A.; Shahabi, H.; Bui, D.T.; Pradhan, B.; Azareh, A.; Rafiei-Sardooi, E.; Samani, A.N.; Melesse, A.M. Groundwater spring potential modelling: Comprising the capability and robustness of three different modeling approaches. *J. Hydrol.* **2018**, *565*, 248–261. [CrossRef]

25. Kim, J.-C.; Jung, H.-S.; Lee, S. Spatial mapping of the groundwater potential of the geum river basin using ensemble models based on remote sensing images. *Remote Sens.* **2019**, *11*, 2285. [CrossRef]

26. Lee, S.; Hyun, Y.; Lee, M.-J. Groundwater potential mapping using data mining models of big data analysis in Goyang-si, South Korea. *Sustainability* **2019**, *11*, 1678. [CrossRef]

27. Lee, S.; Lee, C.-W.; Kim, J.-C. Groundwater productivity potential mapping using logistic regression and boosted tree models: The case of Okcheon city in Korea. In *Advances in Remote Sensing and Geo Informatics Applications*; Springer: Berlin/Heidelberg, Germany, 2019; pp. 305–307.

28. Das, S. Comparison among influencing factor, frequency ratio, and analytical hierarchy process techniques for groundwater potential zonation in Vaitarna basin, Maharashtra, India. *Groundw. Sustain. Dev.* **2019**, *8*, 617–629. [CrossRef]

29. Trabelsi, F.; Lee, S.; Khlifi, S.; Arfaoui, A. Frequency ratio model for mapping groundwater potential zones using gis and remote sensing; Medjerda watershed Tunisia. In *Advances in Sustainable and Environmental Hydrology, Hydrogeology, Hydrochemistry and Water Resources*; Springer: Berlin/Heidelberg, Germany, 2019; pp. 341–345.

30. Arulbalaji, P.; Padmalal, D.; Sreelash, K. GIs and AHp techniques based delineation of groundwater potential zones: A case study from southern Western Ghats, India. *Sci. Rep.* **2019**, *9*, 2082. [CrossRef] [PubMed]

31. Nami, M.; Jaafari, A.; Fallah, M.; Nabiuni, S. Spatial prediction of wildfire probability in the Hyrcanian ecoregion using evidential belief function model and GIS. *Int. J. Environ. Sci. Technol.* **2018**, *15*, 373–384. [CrossRef]

32. Chen, W.; Li, H.; Hou, E.; Wang, S.; Wang, G.; Panahi, M.; Li, T.; Peng, T.; Guo, C.; Niu, C. GIS-based groundwater potential analysis using novel ensemble weights-of-evidence with logistic regression and functional tree models. *Sci. Total Environ.* **2018**, *634*, 853–867. [CrossRef] [PubMed]

33. Naghibi, S.A.; Pourghasemi, H.R.; Abbaspour, K. A comparison between ten advanced and soft computing models for groundwater qanat potential assessment in Iran using R and GIS. *Theor. Appl. Climatol.* **2018**, *131*, 967–984. [CrossRef]

34. Lee, S.; Hong, S.-M.; Jung, H.-S. GIS-based groundwater potential mapping using artificial neural network and support vector machine models: The case of Boryeong city in Korea. *Geocarto Int.* **2018**, *33*, 847–861. [CrossRef]

35. Josephs-Afoko, D.; Godfrey, S.; Campos, L.C. Assessing the performance and robustness of the UNICEF model for groundwater exploration in Ethiopia through application of the analytic hierarchy process, logistic regression and artificial neural networks. *Water Sa* **2018**, *44*, 365–376. [CrossRef]

36. Kumar, A.; Krishna, A.P. Assessment of groundwater potential zones in coal mining impacted hard-rock terrain of India by integrating geospatial and analytic hierarchy process (AHP) approach. *Geocarto Int.* **2018**, *33*, 105–129. [CrossRef]

37. Kordestani, M.D.; Naghibi, S.A.; Hashemi, H.; Ahmadi, K.; Kalantar, B.; Pradhan, B. Groundwater potential mapping using a novel data-mining ensemble model. *Hydrogeol. J.* **2019**, *27*, 211–224. [CrossRef]

38. Miraki, S.; Zanganeh, S.H.; Chapi, K.; Singh, V.P.; Shirzadi, A.; Shahabi, H.; Pham, B.T. Mapping groundwater potential using a novel hybrid intelligence approach. *Water Resour. Manag.* **2019**, *33*, 281–302. [CrossRef]

39. Pham, B.T.; Jaafari, A.; Prakash, I.; Singh, S.K.; Quoc, N.K.; Bui, D.T. Hybrid computational intelligence models for groundwater potential mapping. *Catena* **2019**, *182*, 104101. [CrossRef]

40. Lee, Y.-S.; Park, S.-H.; Jung, H.-S.; Baek, W.-K. Classification of Natural and Artificial Forests from KOMPSAT-3/3A/5 Images Using Artificial Neural Network. *Korean J. Remote Sens.* **2018**, *34*, 1399–1414.

41. Bauer, E.; Kohavi, R. An empirical comparison of voting classification algorithms: Bagging, boosting, and variants. *Mach. Learn.* **1999**, *36*, 105–139. [CrossRef]

42. Gyeonggi Research Institute. *Improvements of the Groundwater Management System in Gyeonggi-do. 2018*; Gyeonggi Research Institute: Gyeonggi-do, South Korea, 2018.

43. K Water. *Groundwater Annual Report. 2017*; Ministry of Environment, Water Policy Coordination Division: Sejong-si, South Korea, 2017.

44. Fallon, A.; Villholth, K.; Conway, D.; Lankford, B.; Ebrahim, G. Agricultural groundwater management strategies and seasonal climate forecasting: Perceptions from Mogwadi (Dendron), Limpopo, South Africa. *J. Water Clim. Chang.* **2019**, *10*, 142–157. [CrossRef]

45. K Water. *Groundwater Basic Survey Report-Yangpyeong-Gun*; Ministry of Environment, Water Policy Coordination Division: Sejong-si, South Korea, 2008.

46. Miralles-Wilhelm, F.; Hejazi, M.; Kim, S.; Yonkofski, C.; Watson, D.; Kyle, P.; Liu, Y.; Vernon, C.; Delgado, A.; Edmonds, J. *Water for Food and Energy Security: An Assessment of the Impacts of Water Scarcity on Agricultural Production and Electricity Generation in the Middle East and North Africa*; World Bank: Washington, DC, USA, 2018.

47. Kalantar, B.; Al-Najjar, H.A.; Pradhan, B.; Saeidi, V.; Halin, A.A.; Ueda, N.; Naghibi, S.A. Optimized conditioning factors using machine learning techniques for groundwater potential mapping. *Water* **2019**, *11*, 1909. [CrossRef]

48. Mogaji, K.A.; Lim, H.S. Development of groundwater favourability map using GIS-based driven data mining models: An approach for effective groundwater resource management. *Geocarto Int.* **2018**, *33*, 397–422. [CrossRef]

49. Naghibi, S.A.; Pourghasemi, H.R. A comparative assessment between three machine learning models and their performance comparison by bivariate and multivariate statistical methods in groundwater potential mapping. *Water Resour. Manag.* **2015**, *29*, 5217–5236. [CrossRef]

50. SAGA-GIS System for Automated Geoscientific Analyses. Available online: www.sagagis.org (accessed on 9 January 2020).

51. Al-Abadi, A.M.; Pradhan, B.; Shahid, S. Prediction of groundwater flowing well zone at An-Najif Province, central Iraq using evidential belief functions model and GIS. *Environ. Monit. Assess.* **2016**, *188*, 549. [CrossRef] [PubMed]

52. Lee, S.; Lee, C.-W. Application of decision-tree model to groundwater productivity-potential mapping. *Sustainability* **2015**, *7*, 13416–13432. [CrossRef]

53. Lee, S.; Lee, S.; Lee, M.-J.; Jung, H.-S. Spatial assessment of urban flood susceptibility using data mining and geographic information System (GIS) tools. *Sustainability* **2018**, *10*, 648. [CrossRef]

54. Lee, S.; Pradhan, B. Landslide hazard mapping at Selangor, Malaysia using frequency ratio and logistic regression models. *Landslides* **2007**, *4*, 33–41. [CrossRef]

55. Pradhan, B. Landslide susceptibility mapping of a catchment area using frequency ratio, fuzzy logic and multivariate logistic regression approaches. *J. Indian Soc. Remote Sens.* **2010**, *38*, 301–320. [CrossRef]
56. Sujatha, E.R.; Rajamanickam, V.; Kumaravel, P.; Saranathan, E. Landslide susceptibility analysis using probabilistic likelihood ratio model—A geospatial-based study. *Arab. J. Geosci.* **2013**, *6*, 429–440. [CrossRef]
57. Lee, S.; Lee, M.-J.; Lee, S. Spatial prediction of urban landslide susceptibility based on topographic factors using boosted trees. *Environ. Earth Sci.* **2018**, *77*, 656. [CrossRef]
58. Bresfelean, V.P. Analysis and predictions on students' behavior using decision trees in Weka environment. In Proceedings of the 29th International Conference on Information Technology Interfaces, Cavtat, Croatia, 25–28 June 2007; pp. 25–28.
59. Zhao, Y.; Zhang, Y. Comparison of decision tree methods for finding active objects. *Adv. Space Res.* **2008**, *41*, 1955–1959. [CrossRef]
60. Breiman, L.; Friedman, J.; Olshen, R.; Stone, C. Classification and regression trees. *Wadsworth Int. Group* **1984**, *37*, 237–251.
61. Kass, G.V. An exploratory technique for investigating large quantities of categorical data. *J. R. Stat. Soc.* **1980**, *29*, 119–127. [CrossRef]
62. Quinlan, J.R. Induction of decision trees. *Mach. Learn.* **1986**, *1*, 81–106. [CrossRef]
63. Quinlan, J.R. *C4. 5: Programs for machine learning*; Elsevier: Amsterdam, The Netherlands, 2014.
64. Friedman, J.H. Stochastic gradient boosting. *Comput. Stat. Data Anal.* **2002**, *38*, 367–378. [CrossRef]
65. Dietterich, T. Overfitting and undercomputing in machine learning. *ACM Comput. Surv.* **1995**, *27*, 326–327. [CrossRef]
66. Schaffer, C. Overfitting avoidance as bias. *Mach. Learn.* **1993**, *10*, 153–178. [CrossRef]
67. Altman, D.G.; Bland, J.M. Diagnostic tests. 1: Sensitivity and specificity. *BMJ* **1994**, *308*, 1552. [CrossRef] [PubMed]
68. Lee, S.; Lee, M.-J.; Jung, H.-S. Data mining approaches for landslide susceptibility mapping in Umyeonsan, Seoul, South Korea. *Appl. Sci.* **2017**, *7*, 683. [CrossRef]
69. Hsu, F.-M.; Lin, Y.-T.; Ho, T.-K. Design and implementation of an intelligent recommendation system for tourist attractions: The integration of EBM model, Bayesian network and Google Maps. *Expert Syst. Appl.* **2012**, *39*, 3257–3264. [CrossRef]
70. Kuncheva, L.I. Classifier ensembles for changing environments. In Proceedings of the International Workshop on Multiple Classifier Systems, Cagliari, Italy, 9–11 June 2004; pp. 1–15.
71. Bui, D.T.; Ho, T.-C.; Pradhan, B.; Pham, B.-T.; Nhu, V.-H.; Revhaug, I. GIS-based modeling of rainfall-induced landslides using data mining-based functional trees classifier with AdaBoost, Bagging, and MultiBoost ensemble frameworks. *Environ. Earth Sci.* **2016**, *75*, 1101.

MDPI
St. Alban-Anlage 66
4052 Basel
Switzerland
Tel. +41 61 683 77 34
Fax +41 61 302 89 18
www.mdpi.com

Remote Sensing Editorial Office
E-mail: remotesensing@mdpi.com
www.mdpi.com/journal/remotesensing